AF147335

Communications
in Computer and Information Science   2620

Series Editors

Gang Li, *School of Information Technology, Deakin University, Burwood, VIC, Australia*
Joaquim Filipe , *Polytechnic Institute of Setúbal, Setúbal, Portugal*
Zhiwei Xu, *Chinese Academy of Sciences, Beijing, China*

Rationale

The CCIS series is devoted to the publication of proceedings of computer science conferences. Its aim is to efficiently disseminate original research results in informatics in printed and electronic form. While the focus is on publication of peer-reviewed full papers presenting mature work, inclusion of reviewed short papers reporting on work in progress is welcome, too. Besides globally relevant meetings with internationally representative program committees guaranteeing a strict peer-reviewing and paper selection process, conferences run by societies or of high regional or national relevance are also considered for publication.

Topics

The topical scope of CCIS spans the entire spectrum of informatics ranging from foundational topics in the theory of computing to information and communications science and technology and a broad variety of interdisciplinary application fields.

Information for Volume Editors and Authors

Publication in CCIS is free of charge. No royalties are paid, however, we offer registered conference participants temporary free access to the online version of the conference proceedings on SpringerLink (http://link.springer.com) by means of an http referrer from the conference website and/or a number of complimentary printed copies, as specified in the official acceptance email of the event.

CCIS proceedings can be published in time for distribution at conferences or as post-proceedings, and delivered in the form of printed books and/or electronically as USBs and/or e-content licenses for accessing proceedings at SpringerLink. Furthermore, CCIS proceedings are included in the CCIS electronic book series hosted in the SpringerLink digital library at http://link.springer.com/bookseries/7899. Conferences publishing in CCIS are allowed to use Online Conference Service (OCS) for managing the whole proceedings lifecycle (from submission and reviewing to preparing for publication) free of charge.

Publication process

The language of publication is exclusively English. Authors publishing in CCIS have to sign the Springer CCIS copyright transfer form, however, they are free to use their material published in CCIS for substantially changed, more elaborate subsequent publications elsewhere. For the preparation of the camera-ready papers/files, authors have to strictly adhere to the Springer CCIS Authors' Instructions and are strongly encouraged to use the CCIS LaTeX style files or templates.

Abstracting/Indexing

CCIS is abstracted/indexed in DBLP, Google Scholar, EI-Compendex, Mathematical Reviews, SCImago, Scopus. CCIS volumes are also submitted for the inclusion in ISI Proceedings.

How to start

To start the evaluation of your proposal for inclusion in the CCIS series, please send an e-mail to ccis@springer.com.

Shehroz S. Khan · Luca Romeo · Ali Abedi
Editors

ArtifiAI for Aging Rehabilitation and Intelligent Assisted Living

8th International Workshop, ARIAL 2025
Held in Conjunction with IJCAI 2025
Montreal, QC, Canada, August 16–22, 2025
Proceedings

 Springer

Editors
Shehroz S. Khan 🆔
KITE Research Institute, University Health
Network
Toronto, Canada

Luca Romeo 🆔
Department of Economics and Law
University of Macerata
Macerata, Italy

Ali Abedi 🆔
KITE Research Institute, University Health
Network
Toronto, ON, Canada

ISSN 1865-0929   ISSN 1865-0937 (electronic)
Communications in Computer and Information Science
ISBN 978-981-95-0567-8   ISBN 978-981-95-0568-5 (eBook)
https://doi.org/10.1007/978-981-95-0568-5

© The Editor(s) (if applicable) and The Author(s), under exclusive license
to Springer Nature Singapore Pte Ltd. 2025

This work is subject to copyright. All rights are solely and exclusively licensed by the Publisher, whether the whole or part of the material is concerned, specifically the rights of translation, reprinting, reuse of illustrations, recitation, broadcasting, reproduction on microfilms or in any other physical way, and transmission or information storage and retrieval, electronic adaptation, computer software, or by similar or dissimilar methodology now known or hereafter developed.
The use of general descriptive names, registered names, trademarks, service marks, etc. in this publication does not imply, even in the absence of a specific statement, that such names are exempt from the relevant protective laws and regulations and therefore free for general use.
The publisher, the authors and the editors are safe to assume that the advice and information in this book are believed to be true and accurate at the date of publication. Neither the publisher nor the authors or the editors give a warranty, expressed or implied, with respect to the material contained herein or for any errors or omissions that may have been made. The publisher remains neutral with regard to jurisdictional claims in published maps and institutional affiliations.

This Springer imprint is published by the registered company Springer Nature Singapore Pte Ltd.
The registered company address is: 152 Beach Road, #21-01/04 Gateway East, Singapore 189721, Singapore

If disposing of this product, please recycle the paper.

Preface

This volume presents the proceedings of the 8th Workshop on AI for Aging, Rehabilitation, and Intelligent Assisted Living (ARIAL), held in conjunction with the International Joint Conference on Artificial Intelligence (IJCAI 2025), in Montreal, Canada, on August 17, 2025.

The ARIAL workshop provides a focused forum for researchers and practitioners working at the intersection of artificial intelligence, aging, and rehabilitation. It aims to foster interdisciplinary dialogue on topics including machine learning for health monitoring, multimodal data collection, telemedicine, sensor-based remote monitoring, privacy-preserving algorithms, and generative AI in elderly care.

The workshop received a total of 14 submissions, all of which were subjected to a double-blind peer review process. Each paper was evaluated by at least two reviewers, ensuring a fair and thorough selection process. After careful consideration, 10 full papers were accepted for inclusion in this volume.

We thank all the authors who submitted their work, the reviewers for their time and expertise, and the broader research community for its support and engagement. We are also grateful to the IJCAI 2025 organizers and Springer for their assistance in producing this volume. We hope that these proceedings will serve as a valuable resource for researchers and practitioners interested in the use of AI to support aging populations and rehabilitation efforts.

June 2025

Shehroz S. Khan
Luca Romeo
Ali Abedi

Organization

General Chair

Shehroz S. Khan   American University of the Middle East, Kuwait
KITE, University Health Network, Canada

Program Committee Chairs

Shehroz S. Khan	American University of the Middle East, Kuwait KITE, University Health Network, Canada
Luca Romeo	University of Macerata, Italy
Ali Abedi	University Health Network, Canada

Program Committee

Amir Ahmad	United Arab Emirates University, UAE
Michele Bernardini	Marche Polytechnic University, Italy
Veselka Boeva	Blekinge Institute of Technology, Sweden
Reza Basiri	University of Toronto, Canada
Andrea Bucci	University of Macerata, Italy
Mariachiara Di Cosmo	Marche Polytechnic University, Italy
Gelareh Hajian	University of Toronto, Canada
Ryan Koh	University of Toronto, Canada
Pratik Kumar Mishra	University of Toronto, Canada
Ali Nawaz	United Arab Emirates University, UAE
Riccardo Rosati	Marche Polytechnic University, Italy

Contents

Early Prediction of Agitation in Community-Dwelling People with Dementia Using Multimodal Sensors and Machine Learning: Benchmarking of State-of-the-Art Techniques

Ali Abedi[1,2]([envelope]) [ORCID], Charlene H. Chu[1,2] [ORCID], and Shehroz S. Khan[1,3] [ORCID]

[1] KITE Research Institute, Toronto Rehabilitation Institute, University Health Network, Toronto, Canada
{ali.abedi,shehroz.khan}@uhn.ca
[2] Lawrence Bloomberg Faculty of Nursing, University of Toronto, Toronto, Canada
[3] College of Engineering and Technology, American University of the Middle East, Egaila, Kuwait
charlene.chu@utoronto.ca

Abstract. Agitation is one of the most common responsive behaviors in people living with dementia, particularly among those residing in community settings without continuous clinical supervision. Timely prediction of agitation can enable early intervention, reduce caregiver burden, and improve the quality of life for both patients and caregivers. This study aimed to develop and benchmark machine learning approaches for the early prediction of agitation in community-dwelling older adults with dementia using multimodal sensor data. A new set of agitation-related contextual features derived from activity data was introduced and employed for agitation prediction. A wide range of machine learning and deep learning models was evaluated across multiple problem formulations, including binary classification for single-timestamp tabular sensor data and multi-timestamp sequential sensor data, as well as anomaly detection for single-timestamp tabular sensor data. The study utilized the Technology Integrated Health Management (TIHM) dataset, the largest publicly available dataset for remote monitoring of people living with dementia, comprising 2,803 days of in-home activity, physiology, and sleep data. The most effective setting involved binary classification of sensor data using the current 6-h timestamp to predict agitation at the subsequent timestamp. Incorporating additional information, such as time of day and agitation history, further improved model performance, with the highest AUC-ROC of 0.9720 and AUC-PR of 0.4320 achieved by the light gradient boosting machine. This work presents the first comprehensive benchmarking of state-of-the-art techniques for agitation prediction in community-based dementia care using privacy-preserving sensor data. The proposed approach demonstrates that accurate, explainable, and efficient agitation prediction is feasible, paving the way for proactive dementia care solutions that support aging in place.

© The Author(s), under exclusive license to Springer Nature Singapore Pte Ltd. 2025
S. S. Khan et al. (Eds.): IJCAI 2025, CCIS 2620, pp. 1–16, 2025.
https://doi.org/10.1007/978-981-95-0568-5_1

Keywords: agitation prediction · early prediction of agitation · multimodal sensors · people living with dementia · machine learning

1   Introduction

Dementia is a progressive neurodegenerative condition that affects millions globally, posing substantial challenges for individuals, families, and healthcare systems [1]. As populations age, the number of people living with dementia (PLwD) continues to rise rapidly [1]. According to the World Health Organization, over 55 million people currently live with dementia worldwide, a figure projected to reach 78 million by 2030 and 139 million by 2050 [2]. This growing prevalence underscores the urgent need for scalable, proactive solutions to manage dementia-related clinical and behavioral challenges, particularly as more PLwD continue to reside in their homes or community settings rather than institutional care [3].

Responsive behaviors in PLwD, such as aggression, wandering, agitation, and resistance to care, often result from unmet needs, environmental stressors, or health issues. Agitation is especially common, presenting as verbal or physical aggression, restlessness, or emotional distress, and affecting up to 70% of PLwD [4]. It can increase caregiver burden, speed cognitive decline, and lead to early institutionalization and poorer quality of life [5,6]. While many studies focus on managing agitation in long-term care [7–10], over 60% of PLwD live in private homes [3]. Still, agitation in these settings is often under-monitored due to a lack of real-time support and early warning tools [11].

Timely prediction of agitation can offer significant benefits in dementia care. Forecasting episodes allows caregivers to implement personalized calming strategies, adjust environments, or intervene early to prevent escalation [12,13]. Early prediction may reduce emergency visits, delay institutionalization, and improve the well-being of both PLwD and caregivers [14]. However, accurately identifying agitation risk remains challenging due to symptom variability, limited monitoring, and a lack of predictive tools in community settings [11].

Monitoring mobility, physiological signals such as blood pressure and heart rate, and sleep patterns can greatly improve agitation management, detection, and prediction, given their established associations with physical activity, autonomic changes, and sleep disruptions in PLwD [15–19]. The emergence of Internet of Things (IoT), wearable sensors, and Artificial Intelligence (AI) offers a promising direction [19–22]. IoT devices such as wearables, smartphones, and ambient sensors can passively collect continuous physiological and behavioral data. These multimodal streams capture daily routines, sleep, and mobility patterns that may precede agitation. Machine learning techniques can analyze this data to uncover hidden patterns and predict agitation risk [10,12,23]. The integration of IoT sensing and AI modeling enables scalable, personalized, and non-invasive agitation monitoring platforms.

While machine learning offers promise for managing agitation in dementia, most research has focused on detecting agitation as it occurs rather than predicting it in advance [9,11,19,23]. This paper distinguishes between detection, reactive and limited, and prediction, which enables timely intervention. Most prior studies have been conducted in institutional settings [7–10], limiting relevance to PLwD living in the community. As a result, a key gap remains in developing privacy-preserving predictive tools tailored for community-dwelling PLwD.

To address these gaps, this study explores the early prediction of agitation in community-dwelling PLwD using multimodal, privacy-preserving sensor data. The proposed method leverages data from mobility and physiological IoT devices to train machine learning and deep learning models capable of forecasting agitation episodes before they occur. This study makes the following contributions:

- A new set of contextual features related to agitation was derived from activity data and utilized for agitation prediction. Additional contextual information, including time of day and agitation history, was incorporated to enhance prediction performance.
- A diverse set of machine learning and deep learning models was evaluated across various problem formulations, including binary classification and anomaly detection for single-timestamp tabular sensor data, and binary classification for multi-timestamp sequential sensor data.
- Extensive experiments were conducted across different settings on a publicly available dataset, with explainability incorporated through analysis of feature importance and contribution.

By predicting agitation before it occurs and using non-intrusive in-home sensors, this study offers a scalable, privacy-conscious framework for supporting PLwD. The results demonstrate the potential of AI-driven remote monitoring to enable proactive dementia care [24].

2   Related Work

This section provides an overview of existing research on agitation detection and prediction in PLwD using multimodal sensors and machine learning and deep learning algorithms.

Sensor-based agitation detection and prediction in PLwD was first reviewed by Khan et al. [23], who demonstrated feasibility but noted a reliance on single-modality actigraphy data. Deters et al. [10] later provided a broader review of institution-based studies, categorizing them under three platforms: Inside-DEM [25], ORCATECH [26], and DAAD [27]. While InsideDEM and ORCAT-ECH supported association analyses between agitation and variables such as mobility, sleep, and heart rate, DAAD enabled predictive model development. These platforms employed various sensors, including video, wearables, and ambient devices, to capture agitation's multifactorial nature. Some of the existing approaches are detailed below.

Khan et al. developed the DAAD platform [27] and conducted one of the earliest studies using multimodal sensors for agitation detection [28]. Data from two PLwD over 2 days were collected via Empatica E4, showing that multimodal inputs outperformed single sensors using a Random Forest (RF) classifier. Spasojevic et al. [29] extended this work with 600 days of data from 17 PLwD, confirming the value of multimodal sensing. To address class imbalance towards non-agitation episodes, Meng et al. [30] introduced a hybrid undersampling and class re-decision method using sequential patterns to improve performance. Iaboni et al. [19] applied feature ranking and developed personalized models, revealing individual differences in feature importance and benefits of individualized detection approaches.

Apart from the studies based on the DAAD platform [27], several works have investigated alternative approaches to agitation detection using vocal features and in-home sensors. Salekin et al. [31] developed a system for identifying different types of verbal agitation, through a combination of acoustic signal processing and text mining with language models, achieving high detection accuracy. Rezvani et al. [32] collected multimodal sensor data from the homes of 96 PLwD using smart plugs, motion sensors, and door sensors, and developed machine learning models to estimate agitation likelihood. To address limited labeled data, they employed a two-stage semi-supervised strategy: a self-supervised transformation learning model generated pseudo-labels from unlabeled data, followed by a Bayesian ensemble trained on labeled samples. This method outperformed Long Short-Term Memory (LSTM) and Support Vector Machine (SVM) baselines in terms of Area Under the Receiver Operating Characteristic Curve (AUC-ROC).

Palermo et al. [11] utilized the Technology Integrated Health Management (TIHM) dataset to develop machine learning models for agitation detection in community-dwelling PLwD, based on statistical features extracted from multimodal sensor data. Among the evaluated models, Logistic Regression (LR) and Gaussian Process classifiers achieved the highest sensitivity and specificity.

Badawi et al. [33] proposed a semi-supervised learning framework for detecting agitation in PLwD using wearable physiological sensor data. Data were collected from 14 participants across three hospitals using the Empatica E4 wristband, which recorded heart rate, acceleration, electrodermal activity, and skin temperature. With only five participants having fully labeled agitation episodes, the study employed a hybrid approach combining variational autoencoders (VAE) for latent feature extraction and a self-training strategy to iteratively generate pseudo-labels for unlabeled data. The best performance was achieved using XGBoost within the self-training framework enhanced by VAE-derived features.

A limited number of studies have focused on predicting agitation before it occurs. Homdee [34] collected environmental data, such as light, temperature, humidity, air pressure, and acoustic noise, from three PLwD over two months. Statistical features were used to train gradient boosting machines and LSTM models, which showed moderate performance, indicating feasibility and room for improvement. Similarly, HekmatiAthar et al. [12] monitored one community-dwelling PLwD over 64 days using sensors that captured ambient noise, illuminance, temperature, pressure, and humidity. An LSTM model was trained to predict agitation, with downsampling applied to address class imbalances towards non-agitation moments.

Based on the reviewed literature, the majority of existing studies have focused primarily on agitation detection, identifying agitation as it occurs [10, 11, 19, 23, 27–33], rather than on its prediction prior to onset. Moreover, much of this research has been conducted in controlled clinical or institutional settings, limiting the generalizability of findings to real-world, home-based environments. Among the relatively few studies that have addressed agitation prediction [12, 34], most were constrained by small sample sizes or limited sensor modalities, thereby restricting their applicability to scalable deployment. In contrast, the present study explores predicting agitation before it occurs. Leveraging a large, real-world, home-based, and multimodal dataset, this work advances the field by addressing key gaps in prediction timing and scalability, with the goal of supporting proactive and personalized dementia care in home settings.

3 Technology Integrated Health Management (TIHM) Dataset

Participants. The TIHM dataset [11] was collected from 56 PLwD residing in home settings. It comprises a total of 2,803 days of data. The duration of data collection per participant ranged from 3 to 90 days, with a mean of 48.91 days and a standard deviation of 23.09 days. Among the participants, 50% were female; 30% were aged 70–80, 47% were 80–90, and 23% were 90–100 years old. Additionally, 89% identified as white, and 25% were living alone.

Data. TIHM captures a rich set of digital biomarkers through a combination of ambient and physiological sensing modalities.

- **Activity**–Continuous passive infrared (PIR) motion sensors were installed across eight locations within participants' homes, including back-door, bathroom, bedroom, fridge-door, front-door, hallway, kitchen, and lounge, capturing ambient motion on an event-driven basis whenever movement occurred
- **Physiology**–In parallel, a suite of physiological sensors was used by participants to measure eight health parameters multiple times per day, including body-temperature, body-weight, diastolic and systolic blood pressure, heart-rate, muscle-mass, total-body-water, and skin-temperature.
- **Sleep**–The dataset also includes detailed sleep data, capturing heart rate and respiratory rate during awake, light, deep, and rapid eye movement sleep episodes.

Out of the 2,803 days of data available in TIHM, the proportions of missing data calculated on a daily basis were 2.89% for activity, 22.90% for physiology, and 70.21% for sleep. Notably, 69.64% of participants had no recorded sleep data.

Label. TIHM provides clinician-verified labels at 6-hour time resolutions for six key clinical events: agitation episodes, abnormal blood pressure (high or low), abnormal body temperature (high or low), dehydration (low body water), abnormal heart rate (high or low), and significant weight changes. The availability of labels aligned with the multimodal sensor data makes the dataset well-suited for supervised machine learning and deep learning model development.

Agitation. Of the 56 participants in TIHM, 27 experienced at least one agitation episode, while 29 did not exhibit any agitation during the monitoring period. In total, 135 agitation episodes were recorded, with the number of episodes per participant ranging from 1 to 33, measured in 6-hour intervals. Across all participants, the mean and standard deviation of agitation episodes were 2.41 and 5.26, respectively. Among those who experienced at least one agitation episode, the mean and standard deviation were 5.00 and 6.71, respectively. Although some participants exhibited multiple agitation episodes within a single day, most episodes were temporally scattered. On a per-participant basis, the mean and standard deviation of the interval between consecutive agitation episodes were 2.43 days and 4.15 days, respectively.

While no agitation episodes were reported before 6:00 a.m., 8.15% occurred between 6:00 a.m. and 12:00 p.m., 58.52% between 12:00 p.m. and 6:00 p.m., and 33.33% between 6:00 p.m. and midnight. The increased occurrence of agitation in the afternoon and evening aligns with what is commonly referred to as the sundowning phenomenon in PLwD [35], during which individuals may experience heightened confusion, restlessness, or distress later in the day.

4   Method

This section presents the proposed pipeline for agitation prediction in PLwD, outlining the preprocessing, feature extraction, problem formulation, and predictive modeling components.

4.1   Preprocessing

Preprocessing involved missing value imputation, standardization, and normalization. As detailed in Sect. 3, the sleep data modality exhibited a high degree of missingness and was therefore excluded from this study, as it was deemed unsuitable for imputation and subsequent predictive modeling [36, 37]. Consequently, activity and physiology data were used for agitation prediction. Missing values were filled using the mean from the training set, applied to both training and test sets. Data were then standardized and normalized, with all transformations fitted on the training data and applied consistently.

4.2   Feature Extraction

The raw activity data consist of motion detection events recorded with second-level accuracy across eight home locations. The first set of activity features includes 32 statistical variables, derived as described in [11]. For each location, hourly activity counts are first computed. Then, for each timestamp (e.g., a 6-hour period), the sum, maximum, mean, and standard deviation of these hourly counts are calculated. Each feature is labeled using the format 'location-count-statistic' (e.g., *hallway-count-sum*).

The second set of activity features comprises eight contextual features, each calculated over a given timestamp (e.g., a 6-hour period) as follows: *total-events*, the total number of detected activities across all monitored locations, reflecting the overall level of movement or environmental interaction; *unique-locations*, the number of distinct locations where activity was detected, indicating the range of spatial engagement within the home; *active-location-ratio*, the proportion of monitored locations that recorded at least one activity event, providing a normalized measure of spatial activity diversity; *private-to-public-ratio*, the ratio of activity events detected in private areas (e.g., bedroom, bathroom) to those in public or shared areas (e.g., kitchen, hallway, lounge), offering insight into the participant's preference for private versus public spaces; *location-entropy*, a measure of the randomness or unpredictability in the participant's spatial behavior. Higher values indicate more evenly distributed activity across locations, while lower values suggest concentration in a limited number of areas; *location-dominance-ratio*, the proportion of activity events that occurred in the most active location, capturing the degree to which behavior was dominated by a single area; *back-and-forth-count*, the number of instances where a participant returned to a previously visited location after briefly moving to another, potentially indicating restlessness or repetitive movement patterns; *num-transitions*, the total number of transitions between different locations, serving as a measure of mobility and spatial movement complexity.

For each timestamp (e.g., a 6-hour window) in TIHM, there are typically a few, sometimes only one, recorded values for each of the eight physiological measurements. The corresponding eight physiology features are computed as the mean of all available measurements within each timestamp. Therefore, a total of 48 features are extracted from the activity and physiology data at each timestamp, comprising 32 statistical activity features, 8 contextual activity features, and 8 physiology features.

4.3   Agitation Prediction Problem Formulation

This section presents the proposed formulations for agitation prediction, including binary classification using tabular and sequential sensor data, as well as anomaly detection using tabular sensor data.

Binary Classification. The goal is to predict agitation at the next timestamp $(t + 1)$, using sensor data from the previous n timestamps $(t - (n - 1), \ldots, t)$, based on the cyclic nature of dementia symptoms [38–40]. For $n = 1$, each sample is a single feature vector at time t, forming a tabular dataset suitable for standard machine learning models. For $n > 1$, each sample becomes a sequence of n feature vectors, requiring sequential models designed to handle temporal dependencies.

Anomaly Prediction. While agitation detection has been widely explored as an anomaly detection task [9,41], this study examines agitation prediction as an anomaly prediction problem. Here, a sequence of n past timestamps is labeled *normal* if agitation does not occur at $t+1$, and *anomalous* otherwise. Models are trained exclusively on normal samples to learn the underlying distribution of non-agitated behavior. During inference, deviations from this learned distribution indicate a higher likelihood of agitation in the next timestamp.

4.4 Predictive Modeling

A range of machine learning and deep learning models were employed for agitation prediction, based on the problem formulations described in Subsect. 4.3. The models designed for single-timestamp tabular data included a Transformer-based model for tabular data, named Tabular Prior-data Fitted Network (TabPFN) [42] as well as gradient boosting decision tree algorithms, including Gradient Boosting (GB) classifier [43] and Light Gradient Boosting Machine (LightGBM) [44]. In addition, traditional machine learning classifiers were evaluated [45], including LR and Naive Bayes (NB). The models employed for multi-timestamp sequential data included the Transformer architecture [46] and ROCKET (RandOm Convolutional KErnel Transform) [47]. Finally, the models used for anomaly prediction on single-timestamp tabular data included One-Class SVM, Isolation Forest (IF), and Local Outlier Factor (LOF) [48].

Due to the highly imbalanced distribution of data samples in the dataset, with agitation, the class of interest, being the minority, class imbalance was addressed using weighted loss functions [49] and the Synthetic Minority Over-sampling Technique (SMOTE) [50]. The weights in the loss functions were set proportional to the class distribution, ensuring greater penalization for misclassifying minority class instances.

5   Experiments

This section presents the experimental settings and results of the proposed method across different settings for agitation prediction in PLwD.

Agitation Prediction as Binary Classification–Tabular Sensor Data. Table 1 presents results for agitation prediction as a binary classification problem, identifying whether agitation occurs at $t + 1$ using data from t ($n = 1$, per Subsect. 4.3), based on 6-hour samples from TIHM. Models were trained on statistical and contextual activity and physiology features, with and without weighted loss functions, and evaluated using both 5-fold CV and leave-one-participant-out (LOPO) CV settings.

Table 1 shows that both AUC-ROC and AUC-PR values significantly surpass the baseline performance of a random binary classifier. Weighted models generally outperform their unweighted counterparts. As expected, LOPO results are lower than those from 5-fold CV but remain comparable, indicating generalization to unseen participants. The Transformer (TabPFN) and LR models achieve the highest AUC-ROC across different settings, while NB attains the highest AUC-PR. Although NB and LR detect the most agitation episodes, as reflected in their high sensitivity, they also produce more false positives, resulting in lower specificity. Due to the relatively better performance of models trained with weighted loss functions, all subsequent experiments employed this setting.

Agitation Prediction as Binary Classification–Sequential Sensor Data.

Fig. 1 presents binary classification results using data from the n most recent timestamps as sequential data samples from TIHM. Models include a Transformer encoder and a combination of ROCKET [47] and GB [43] with a weighted

Table 1. Agitation prediction formulated as a binary classification problem using tabular sensor data under 5-fold and leave-oneparticipant-out (LOPO) cross-validation (CV). Models marked with a superscript 'W' were trained with a weighted loss function.

CV	Model	AUC-ROC	AUC-PR	Accuracy	F1-score	Sensitivity	Specificity	Gmean
	Gradient Boosting	0.8839	0.1790	0.9875	0	0	1	0
	LightGBM	0.9119	0.1805	0.9874	0	0	0.9999	0
	Logistic Regression	0.9091	0.2007	0.9870	0.0541	0.0296	0.9992	0.1712
	Naïve Bayes	0.8929	0.2844	0.8061	0.1021	0.8815	0.8052	0.8425
5-fold	Transformer	0.9151	0.2037	0.9873	0.0144	0.0074	0.9997	0.0860
	Gradient Boosting ^{W}	0.893	0.176	0.9087	0.1826	0.8148	0.9099	0.8610
	LightGBM ^{W}	0.9099	0.1897	0.9857	0.1895	0.1333	0.9965	0.3645
	Logistic Regression ^{W}	0.9162	0.2105	0.8782	0.1468	0.8370	0.8787	0.8576
	Naïve Bayes ^{W}	0.8922	0.3088	0.7576	0.0853	0.9037	0.7558	0.8265
	Gradient Boosting	0.8105	0.1345	0.9875	0	0	1	0
	LightGBM	0.8573	0.1509	0.9875	0	0	1	0
	Logistic Regression	0.8931	0.1964	0.9871	0.0795	0.0444	0.9991	0.2106
	Naïve Bayes	0.8851	0.2732	0.7973	0.0981	0.8815	0.7962	0.8378
LOPO	Transformer	0.8932	0.2002	0.9872	0.0282	0.0148	0.9995	0.1216
	Gradient Boosting ^{W}	0.8588	0.1580	0.9123	0.1831	0.7852	0.9139	0.8471
	LightGBM ^{W}	0.8712	0.1762	0.9854	0.0814	0.0519	0.9972	0.2275
	Logistic Regression ^{W}	0.8900	0.2026	0.8750	0.1435	0.8370	0.8755	0.8560
	Naïve Bayes ^{W}	0.8847	0.3000	0.7502	0.0830	0.9037	0.7483	0.8223

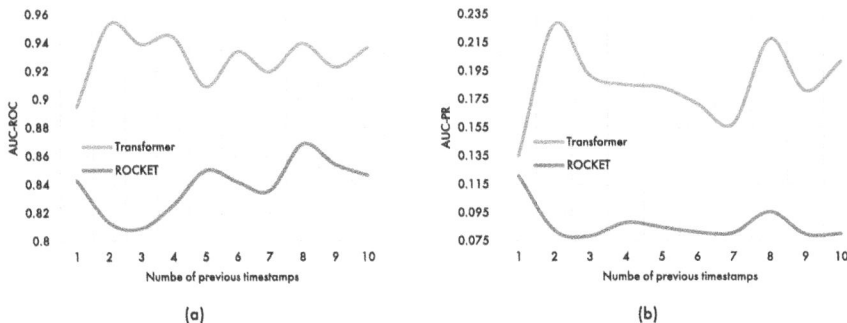

Fig. 1. (a) AUC-ROC and (b) AUC-PR for agitation prediction framed as a binary classification problem using sequential sensor data under 5-fold cross-validation. The Transformer consistently outperforms ROCKET, achieving peak performance with an AUCROC of 0.9531 and AUC-PR of 0.2277 when using the two most recent timestamps to predict agitation at the next timestamp.

Table 2. Agitation prediction formulated as an anomaly prediction problem using tabular sensor data under 5-fold and leave-oneparticipant-out (LOPO) cross-validation (CV). Models marked with a superscript 'W' were trained with a weighted loss function.

CV	Model	AUC-ROC	AUC-PR	Accuracy	F1-score	Sensitivity	Specificity	Gmean
	One-class SVM	0.7181	0.4393	0.5047	0.0416	0.8593	0.5002	0.6556
5-fold	Isolation Forest	0.8419	0.0783	0.9273	0.1552	0.5333	0.9323	0.7052
	Local Outlier Factor	0.6200	0.0170	0.9499	0.0146	0.0296	0.9615	0.1688
	One-class SVM	0.7077	0.4387	0.4923	0.0410	0.8667	0.4876	0.6508
LOPO	Isolation Forest	0.8376	0.0675	0.9129	0.1296	0.5185	0.9179	0.6896
	Local Outlier Factor	0.5927	0.0146	0.8787	0.0195	0.0963	0.8886	0.2925

loss function, trained on statistical and contextual activity and physiology features under 5-fold CV. The Transformer consistently outperforms ROCKET, achieving the highest AUC-ROC and AUC-PR at $n = 2$. When more than two past timestamps are used, the Transformer's agitation prediction performance declines.

Agitation Prediction as Anomaly Prediction. Table 2 presents results for agitation prediction framed as an anomaly prediction problem, using data from the current timestamp ($n = 1$) with 6-hour samples from TIHM, evaluated under 5-fold and LOPO CV. Models were trained only on normal data (non-agitation at the next timestamp). Compared to binary classification (Table 1), anomaly prediction yields lower AUC-ROC overall, though One-class SVM achieves higher AUC-PR than the classification models. These results suggest that while anomaly prediction may struggle with overall discrimination, it can be more effective in identifying rare agitation cases.

5.1   Temporal Resolution

Figure 2 shows agitation prediction performance, formulated as binary classification of tabular data across 6-hour, 12-hour, and 24-hour temporal resolutions. The number of positive cases varies slightly by resolution, with samples containing multiple agitation episodes labeled the same as those with only one. Longer intervals result in fewer total samples, reducing available negative cases for training and testing. The 6-hour setting yields the highest AUC-PR and AUC-ROC, followed by 12-hour and then 24-hour, indicating improved prediction accuracy with shorter temporal windows. These findings underscore the clinical value of higher-frequency monitoring for timely and reliable agitation prediction.

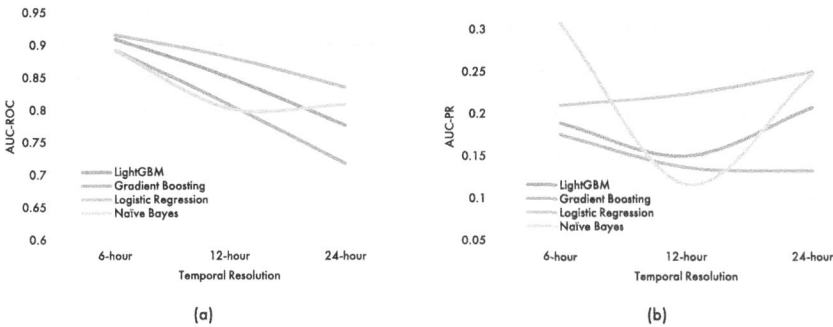

Fig. 2. (a) AUC-ROC and (b) AUC-PR for agitation prediction framed as a binary classification problem using tabular sensor data under 5-fold cross-validation for 6-hour, 12-hour, and 24-hour temporal resolutions.

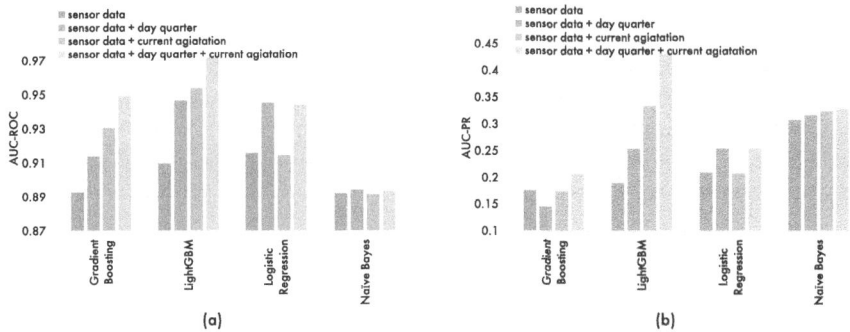

Fig. 3. (a) AUC-ROC and (b) AUC-PR for agitation prediction framed as a binary classification problem using tabular sensor data under 5-fold cross-validation, following the incorporation of additional information: day quarter and current agitation status.

5.2   Incorporating Additional Information

As noted in Sect. 3, agitation episodes in TIHM were unevenly distributed across the day, with a higher concentration in the afternoon and evening. This pattern reflects the well-known sundowning phenomenon in PLwD [35], motivating the inclusion of day-quarter information (encoded as an integer) as an additional input to models alongside activity and physiology features. Temporal analysis also showed that 86.67% of agitation episodes occurred without agitation in the previous timestamp, while only 13.33% were preceded by an agitation. This motivated the inclusion of a binary feature indicating current agitation to support the prediction of future agitation. Importantly, incorporating day-quarter and current-timestamp agitation as input features does not introduce information leakage, as both are derived solely from the current or past timestamps and exclude any future data.

Figure 3 presents the results of the proposed method under 5-fold CV with incremental inclusion of additional contextual information. Incorporating each piece of information individually, and especially both together, leads to substantial improvements in AUC-ROC and AUC-PR across all settings, reaching up to 0.9720 and 0.4320, respectively, for LightGBM. These results highlight the value of integrating contextual cues to enhance agitation prediction performance.

5.3   Feature Importance and Contribution

SHapley Additive exPlanations (SHAP) [51] was used to evaluate the contribution of statistical, contextual activity, and physiology features to agitation prediction. LightGBM, selected for its strong performance in earlier experiments, served as the base model under 5-fold CV. Figure 4 shows the SHAP summary plot of the top 24 features, highlighting both their importance and the effect of low vs. high values on predictions. Features are ranked by impact, with the top feature being a statistical activity measure–hallway-count-std (see Sect. 4.2). The predominance of high values (in red) on the positive SHAP axis suggests that greater hallway movement variability increases the likelihood of agitation in the next timestamp. Similar interpretations apply to other features based on SHAP value distribution and color. Notably, several proposed contextual features also rank among the most influential.

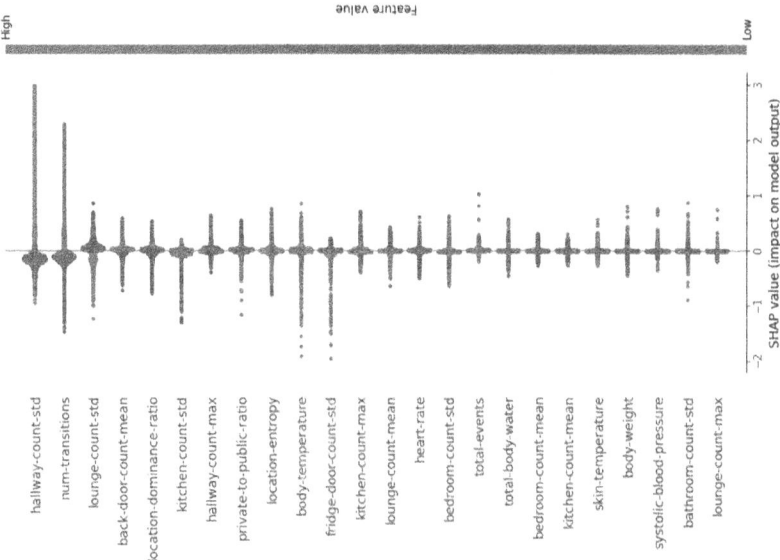

Fig. 4. SHapley Additive exPlanations (SHAP) summary plot for the first 24 most important features for agitation prediction.

6   Conclusion

This study investigated early prediction of agitation episodes in community-dwelling PLwD using multimodal sensor data and machine learning. Leveraging the TIHM dataset, which contains continuous in-home activity and physiological data, multiple problem formulations were explored. Novel contextual features derived from activity data were introduced for prediction. The binary classification model using the current 6-hour timestamp consistently achieved the best performance, further enhanced by incorporating time-of-day and current agitation status information. A key limitation of this study is the limited number of agitation episodes for model training, which may impact generalizability and lead to lower true positive rates. Although the TIHM dataset provides rich multimodal data, it represents a specific cohort of community-dwelling older adults with dementia, potentially limiting applicability to other settings or populations. The use of fixed 6-, 12-, and 24-hour timestamps may also miss finer temporal patterns. Future work should explore larger, more diverse datasets and adaptive temporal models to enhance prediction robustness and relevance. In conclusion, this study demonstrates the feasibility of accurate and explainable agitation prediction in community-dwelling PLwD using privacy-preserving, non-invasive multimodal sensor data. The results hold significant promise for proactive dementia care by supporting timely interventions and alleviating caregiver burden.

Acknowledgment. This work was supported by Longitude Prize on Dementia, UK, and the Alzheimer Society of Canada.

References

1. Livingston, G., et al.: Dementia prevention, intervention, and care: 2024 report of the lancet standing commission. The Lancet **404**(10452), 572–628 (2024)
2. World Health Organization, "Dementia," 2023. https://www.who.int/news-room/fact-sheets/detail/dementia. Accessed 16 Apr 2025
3. Canadian Institute for Health Information, "Dementia in home and community care," (2018). https://www.cihi.ca/en/dementia-in-canada/dementia-care-across-the-health-system/dementia-in-home-and-community-care. Accessed 16 Apr 2025
4. Carrarini, C., et al.: Agitation and dementia: prevention and treatment strategies in acute and chronic conditions. Front. Neurol. **12**, 644317 (2021)
5. Cloutier, M., et al.: Institutionalization risk and costs associated with agitation in Alzheimer's disease. Alzheimer's Dementia: Trans. Res. Clin. Intervent. **5**, 851–861 (2019)
6. Schein, J., et al.: The impact of agitation in dementia on caregivers: a real-world survey. J. Alzheimer s Disease **88**(2), 663–677 (2022)
7. Wong, B., Ismail, Z., Watt, J., Holroyd-Leduc, J., Goodarzi, Z.: Barriers and facilitators to care for agitation and/or aggression among persons living with dementia in long-term care. BMC Geriatr. **24**(1), 330 (2024)

8. Khan, S.S., Mishra, P.K., Ye, B., Patel, S., Newman, K., Mihailidis, A., Iaboni, A.: A novel multi-modal sensor dataset and benchmark to detect agitation in people living with dementia in a residential care setting. ACM Trans. Comput. Healthcare **6**(3), 1–19 (2025). https://doi.org/10.1145/3720550

9. Mishra, P.K., Iaboni, A., Ye, B., Newman, K., Mihailidis, A., Khan, S.S.: Privacy-protecting behaviours of risk detection in people with dementia using videos. Biomed. Eng. Online **22**(1), 4 (2023)

10. Deters, J.K., Janus, S., Silva, J., Wörtche, H.J., Zuidema, S.U.: Sensor-based agitation prediction in institutionalized people with dementia a systematic review. Pervasive Mob. Comput. **98**, 101876 (2024)

11. Palermo, F., et al.: TIHM: An open dataset for remote healthcare monitoring in dementia. Scientific data **10**(1), 606 (2023)

12. HekmatiAthar, S., Goins, H., Samuel, R., Byfield, G., Anwar, M.: Data-driven forecasting of agitation for persons with dementia: a deep learning-based approach. SN Comput. Sci. **2**(4), 326 (2021)

13. Anderson, M.S., Bankole, A., Homdee, N., Mitchell, B.A., Byfield, G.E., Lach, J.: Dementia caregiver experiences and recommendations for using the behavioral and environmental sensing and intervention system at home: usability and acceptability study. JMIR Aging **4**(4), e30353 (2021)

14. Gitlin, L.N., Kales, H.C., Lyketsos, C.G.: Nonpharmacologic management of behavioral symptoms in dementia. JAMA **308**(19), 2020–2029 (2012)

15. Cheung, J.C.-W., So, B.P.-H., Ho, K., Wong, D.W.-C., Lam, A.H.-F., Cheung, D.: Wrist accelerometry for monitoring dementia agitation behaviour in clinical settings: A scoping review. Front. Psych. **13**, 913213 (2022)

16. Liu, K.Y., et al.: Heart rate variability and risk of agitation in Alzheimer's disease: the atherosclerosis risk in communities study. Brain Communications **5**(6) (2023). https://doi.org/10.1093/braincomms/fcad269

17. De Heus, R.A., et al.: Association between blood pressure variability with dementia and cognitive impairment: a systematic review and meta-analysis. Hypertension **78**(5), 1478–1489 (2021)

18. Brown, D.T., Westbury, J.L., Schüz, B.: Sleep and agitation in nursing home residents with and without dementia. Int. Psychogeriatr. **27**(12), 1945–1955 (2015)

19. Isboni, A., et al.: Wearable multimodal sensors for the detection of behavioral and psychological symptoms of dementia using personalized machine learning models. Alzheimer's & Dementia: Diagnosis, Assessment Disease Monitoring **14**(1) (2022). https://doi.org/10.1002/dad2.12305

20. Abedi, A., Colella, T.J., Pakosh, M., Khan, S.S.: Artificial intelligence-driven virtual rehabilitation for people living in the community: a scoping review. NPJ Digital Med. **7**(1), 25 (2024)

21. Abedi, A., Dayyani, F., Chu, C., Khan, S.S.: "Maison-multimodal ai-based sensor platform for older individuals. In: 2022 IEEE International Conference on Data Mining Workshops (ICDMW). IEEE, 2022, pp. 238–242 (2022)

22. Sheikhtaheri, A., Sabermahani, F.: Applications and outcomes of internet of things for patients with Alzheimer s disease/dementia: A scoping review. Biomed. Res. Int. **2022**(1), 6274185 (2022)

23. Khan, S.S., Ye, B., Taati, B., Mihailidis, A.: Detecting agitation and aggression in people with dementia using sensors a systematic review. Alzheimer's Dementia **14**(6), 824–832 (2018)

24. Waldman, S.A., Terzic, A.: Healthcare evolves from reactive to proactive. Clin. Pharmacol. Ther. **105**(1), 10 (2019)

25. Teipel, S., et al.: Multidimensional assessment of challenging behaviors in advanced stages of dementia in nursing homes the insidedem framework. Alzheimer's & Dementia: Diagnosis, Assessment Disease Monitoring **8**, 36–44 (2017)
26. Kaye, J., et al.: Methodology for establishing a community-wide life laboratory for capturing unobtrusive and continuous remote activity and health data. J. Visual. Exper. (137) (2018). https://doi.org/10.3791/56942-v
27. Khan, S.S.: "Daad: A framework for detecting agitation and aggression in people living with dementia using a novel multi-modal sensor network." In: 2017 IEEE International Conference on Data Mining Workshops (ICDMW). IEEE, 2017, pp. 703–710 (2017)
28. Khan, S.S., et al.: Agitation detection in people living with dementia using multi-modal sensors. In: 41st Annual International Conference of the IEEE Engineering in Medicine and Biology Society (EMBC). IEEE **2019**, 3588–3591 (2019)
29. Spasojevic, S., et al.: A pilot study to detect agitation in people living with dementia using multi-modal sensors. J. Healthcare Inform. Res. **5**(3), 342–358 (2021)
30. Meng, Z., et al.: Undersampling and cumulative class re-decision methods to improve detection of agitation in people with Dementia. Biomed. Eng. Lett. **14**(1), 69–78 (2024)
31. Salekin, A., Wang, H., Williams, K., Stankovic, J.: Dave: detecting agitated vocal events. In: 2017 IEEE/ACM International Conference on Connected Health: Applications, Systems and Engineering Technologies (CHASE). IEEE, 2017, pp. 157–166 (2017)
32. Rezvani, R., Kouchaki, S., Nilforooshan, R., Sharp, D.J., Barnaghi, P.: Semi-supervised learning for identifying the likelihood of agitation in people with dementia. *arXiv preprint*arXiv:2105.10398, (2021)
33. Badawi, A., Elmoghazy, S., Choudhury, S., Elgazzar, K., Burhan, A.: Leveraging self-training and variational autoencoder for agitation detection in people with dementia using wearable sensors. arXiv preprint arXiv:2412.19254, (2024)
34. Homdee, N.: Prediction of dementia-related agitation using multivariate ambient environmental time-series data. arXiv preprint arXiv:2002.07237, (2020)
35. Canevelli, M., et al.: Sundowning in dementia: clinical relevance, pathophysiological determinants, and therapeutic approaches. Front. Med. **3**, 73 (2016)
36. Madley-Dowd, P., Hughes, R., Tilling, K., Heron, J.: The proportion of missing data should not be used to guide decisions on multiple imputation. J. Clin. Epidemiol. **110**, 63–73 (2019)
37. Junaid, K., Kiran, T., Gupta, M., Kishore, K., Siwatch, S.: How much missing data is too much to impute for longitudinal health indicators? a preliminary guideline for the choice of the extent of missing proportion to impute with multiple imputation by chained equations. Popul. Health Metrics **23**(1), 2 (2025)
38. Rak-Areekul, T., Likijaroen, Y., Hiransuthikul, A., Booncharoen, K., Sarutikriangkri, Y.: Prevalence of people living with dementia with cyclic temporal pattern of behavioral and psychological symptoms of dementia in a tertiary memory clinic: A descriptive diary-based study. Alzheimer's & Dementia **20**, e085858 (2024)
39. Bachman, D., Rabins, P.: sundowning and other temporally associated agitation states in dementia patients. Annu. Rev. Med. **57**(1), 499–511 (2006)
40. Cohen-Mansfield, J.: Temporal patterns of agitation in dementia. Am. J. Geriatr. Psychiatry **15**(5), 395–405 (2007)
41. Khan, S.S., Mishra, P.K., Javed, N., Ye, B., Newman, K., Mihailidis, A., Iaboni, A.: Unsupervised deep learning to detect agitation from videos in people with dementia. IEEE Access **10**, 10349–10358 (2022). https://doi.org/10.1109/ACCESS.2022.3143990

42. Hollmann, N., et al.: Accurate predictions on small data with a tabular foundation model. Nature **637**(8045), 319–326 (2025)
43. Friedman, J.H.: Greedy function approximation: a gradient boosting machine. Ann. Stat. 1189–1232 (2001)
44. Ke, G., et al.: Lightgbm: A highly efficient gradient boosting decision tree. In: Advances in Neural Information Processing Systems, vol. 30 (2017)
45. Pranckevičius, T., Marcinkevičius, V.: Comparison of Naive Bayes, random forest, decision tree, support vector machines, and logistic regression classifiers for text reviews classification. Baltic J. Modern Comput. **5**(2), 221 (2017)
46. Vaswani, A., et al.: Attention is all you need. In: Advances in Neural Information Processing Systems, vol. 30 (2017)
47. Dempster, A., Petitjean, F., Webb, G.I.: Rocket: exceptionally fast and accurate time series classification using random convolutional kernels. Data Min. Knowl. Disc. **34**(5), 1454–1495 (2020)
48. Ahmad, A., Khan, S.: One-class classification with node embedding type features. Inform. Sci. 122204 (2025)
49. Cui, Y., Jia, M., Lin, T.-Y., Song, Y., Belongie, S.: Class-balanced loss based on effective number of samples. In: Proceedings of the IEEE/CVF Conference on Computer Vision and Pattern Recognition, pp. 9268–9277 (2019)
50. Chawla, N.V., Bowyer, K.W., Hall, L.O., Kegelmeyer, W.P.: Smote: synthetic minority over-sampling technique. J. Artif. Intell. Res. **16**, 321–357 (2002)
51. Lundberg, S.M., Lee, S.-I.: A unified approach to interpreting model predictions. In: Advances in neural information processing systems, vol. 30 (2017)

Conversational AI for Cognitive, Emotional, and Social Engagement of Elderly Persons: A Large Language Model-Based Framework

Ali Nawaz⬩ and Amir Ahmad(✉)⬩

College of Information Technology and Big Data Analytics Center, United Arab
Emirates University, P.O. Box 15551, Al Ain, United Arab Emirates
amirahmad@uaeu.ac.ae

Abstract. The global rise in the elderly population presents critical
challenges in ensuring mental well-being, addressing social isolation,
and enabling early detection of cognitive decline. The paper proposes
a novel framework leveraging Large Language Models (LLMs) to provide empathetic and intelligent friendship to elderly individuals while
monitoring their cognitive and emotional health. Our proposed system
integrates a dialogue manager, memory-enhanced interaction, embedded
cognitive assessments, sentiment analysis, and a caregiver dashboard.
The framework supports personalized engagement via natural conversation, enabling non-invasive, culturally adaptable, and privacy-preserving
interactions. It will also include passive interaction using group discussions, incorporating family-related information extracted and updated
from social media platforms such as Facebook, Instagram, and WhatsApp etc. The proposed approach holds promise for aiding healthcare
providers in the early detection and continuous monitoring of age-related
mental conditions such as Mild Cognitive Impairment (MCI), social isolation, and depression.

1 Introduction

As the global aging population continues to expand [1], there is a corresponding increase in risks related to social isolation, cognitive decline, and emotional
health disorders [2]. Social isolation among elderly people is not only a psychological concern but also a major risk factor for physical illnesses such as
cardiovascular diseases, weakened immune function, and increased mortality [3].
Addressing these issues requires scalable and intelligent systems that go beyond
traditional clinical visits [4].

Conventional tools for assessing cognitive dysfunction often require in-person
evaluations in clinical settings, limiting their scalability and continuity [5]. Additionally, these tools are often detached from the day-to-day realities and behaviors of aging individuals [6]. The use of Large Language Models (LLMs) presents
a transformative opportunity to bridge this gap by providing context-aware,
empathetic, and interactive virtual companions [7]. These models can simulate

© The Author(s), under exclusive license to Springer Nature Singapore Pte Ltd. 2025
S. S. Khan et al. (Eds.): IJCAI 2025, CCIS 2620, pp. 17–28, 2025.
https://doi.org/10.1007/978-981-95-0568-5_2

human-like conversations, monitor behavior patterns, and deliver embedded cognitive and emotional health checks seamlessly within natural dialogues [8].

Recent advancements in artificial intelligence, particularly transformer-based architectures such as Generative Pretrained Transformer 4 (GPT-4) [9] and Large Language Model Meta AI (LLaMA2) [10], have demonstrated superior performance in understanding context, sentiment, and user intent. Integrating these models into conversational platforms can foster meaningful engagement with aging users, combat social isolation, and support mental well-being [2].

This paper introduces a comprehensive system that combines the power of LLMs with embedded cognitive and emotional assessment tools to address the multifaceted challenges faced by elderly people. By leveraging data from everyday conversations and potentially from social media interactions, our system aims to provide early indicators of mental health decline and promote emotional resilience through companionship and monitoring. Additionally, the integration of family group data enables the LLM to relay meaningful social updates to elderly users, helping them stay connected and emotionally engaged even if they are not active participants in those conversations, thus playing a critical role in reducing social isolation.

The rest of the paper is organized as follows: Sect. 2 presents related work, system architecture is presented in Sect. 3, followed by Sect. 4 related to cognitive and emotional assessment, and then followed by Sects. 5 and 6 related to data privacy and evaluation plan respectively. Finally, Sect. 7 concludes the paper.

2   Related Work

Previous solutions targeting elderly care and cognitive monitoring have spanned a variety of approaches. Robotic companions [11] have been employed to provide emotional support to elderly individuals. While these systems offer physical presence and limited interaction, they lack linguistic depth and contextual adaptability.

Rule-based chatbots designed for dementia care or medication reminders have also been explored. However, these systems are often rigid, incapable of adapting to dynamic conversations, and fail to build long-term user profiles [12]. Gamified cognitive training tools like Lumosity and BrainHQ aim to enhance cognitive abilities but do not address emotional well-being or social interaction [13].

In contrast, LLMs have emerged as a powerful alternative due to their capacity to generate human-like language, recall context, and adapt to diverse conversational scenarios [14]. AI-driven companionship, such as Replika, has shown that conversational agents can foster emotional bonds with users [15]. However, these applications are not specifically designed for aging populations or integrated with cognitive testing frameworks.

A few recent studies have begun exploring the use of conversational AI in mental health and elderly care [16]. ChatGPT has been studied for therapeutic interactions [17], while specialized platforms like Ellie by USC Institute for Creative Technologies have been used for Post-traumatic stress disorder (PTSD) assessment [18]. Despite their promise, these systems either lack personalization for elderly users or fall short in integrating cognitive and emotional analytics, or are not updated with dynamic social inputs such as family group discussions. As a result, they miss crucial opportunities to reinforce social connectedness and emotional well-being through socially contextualized dialogue.

Our work distinguishes itself by proposing a unified framework that not only engages elderly individuals through LLM-powered dialogues but also embeds cognitive stimuli and emotional analysis into the flow of conversation. This integrated approach is uniquely positioned to address social isolation and cognitive health monitoring in aging populations using scalable, ethical, and culturally sensitive AI technology.

3   System Architecture

The proposed framework highlighted in Fig. 1 is composed of modular and interlinked components designed to deliver real-time, personalized, and clinically relevant interactions for elderly users.

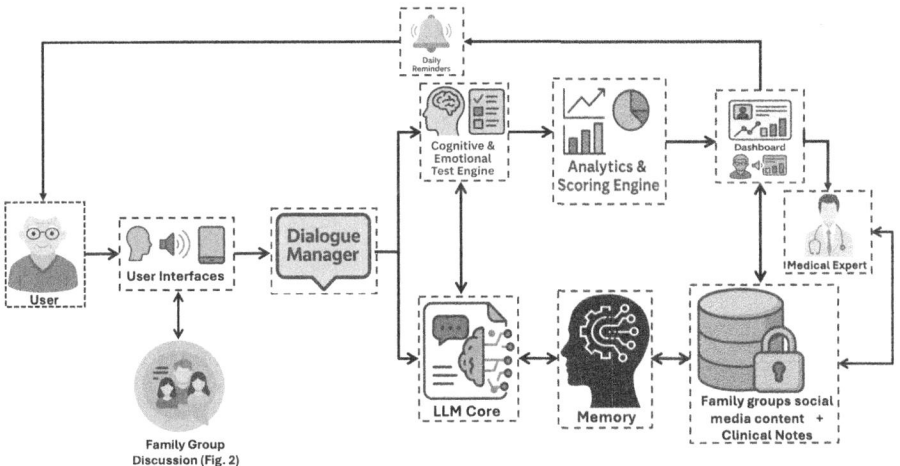

Fig. 1. Schematic diagram illustrating the proposed framework

The modules are:

– User Interface: A multimodal interface accessible via smartphones, tablets, or smart home devices [19]. It supports both text-based and voice-based interactions, enabling usability for elderly individuals with varying levels of technological literacy. Features like large fonts, contrast settings, and voice commands ensure accessibility.

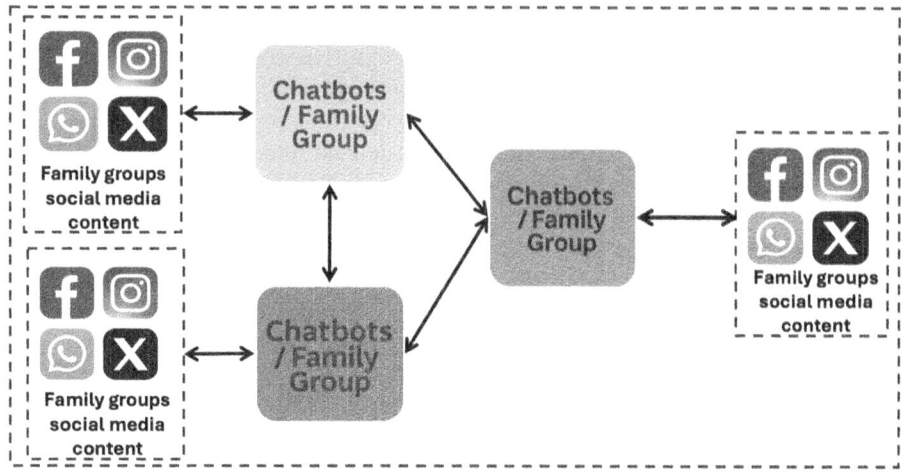

Fig. 2. LLM chatbots connected to family group discussions via social media (Facebook, Instagram, WhatsApp, X etc.) to deliver updates to the elderly users to combat social isolation

- Dialogue Manager: Acts as the central server, determining the intent behind each user input. It switches seamlessly between casual conversation and embedded cognitive/emotional assessments, maintaining flow and empathy while ensuring necessary evaluations are carried out [20].
- LLM Core: The heart of the system, consisting of a fine-tuned transformer model such as GPT-4 [9] or LLaMA2 [10]. This component is specifically tailored to exhibit elderly-friendly dialogue behaviors, including simplified language, cultural idioms, memory recall, and reinforcement of identity.
- Cognitive and Emotional Testing Engine: Integrates standard clinical instruments like MMSE (Mini-Mental State Examination), MoCA (Montreal Cognitive Assessment) [21], and the Geriatric Depression Scale into the natural flow of conversation. This engine dynamically adjusts question phrasing and sequence to match the user's cognitive state and comfort level.
- Analytics Engine: Continuously processes linguistic, cognitive, and emotional signals from the conversation. It computes longitudinal trends and anomaly scores indicating early cognitive deterioration or emotional distress.
- Memory Module: A long-term interaction memory that stores facts about the user such as family members, hobbies, routines, and past conversations. It enhances personalization and continuity, allowing the LLM to reference familiar elements and build a deeper relationship with the user over time.
- Database: The database module is a central repository for structured and unstructured data gathered during interactions and from external sources. Historical interaction data from the previous assessments and caregiver notes are stored to construct a comprehensive longitudinal profile. Additionally, family group discussions through social media platforms for training LLMs

are also stored. This allows the system to identify engagement gaps, such as a lack of participation in group conversations, and to generate personalized prompts or interventions that encourage social inclusion and monitor behavioral changes over time

- Group discussion: As shown in Fig. 2, an elderly user interacts with an LLM-powered chatbot that not only engages in direct conversation but also connects to family group discussions through secure databases. The LLM serves as a personalized companion, delivering updates from family group chats such as shared events or messages while responding empathetically to the user's mood and interests. Even if the elder is not active in those groups, the chatbot helps them feel involved by summarizing key moments or prompting gentle social engagement. With appropriate consent, family social media group chat content is sourced from popular social media platforms such as Facebook, Instagram, and WhatsApp, etc. This interaction fosters emotional connection and combats social isolation through consistent, and meaningful dialogue.
- Dashboard: A secure clinical dashboard accessible to authorized caregivers and clinicians. It presents aggregated visualizations, cognitive test scores, sentiment trends, and potential alerts based on predefined thresholds. The portal also includes the ability to annotate sessions and integrate notes into Electronic Health Records (EHR).

Table 1 highlights the technical tools used for developing each component.

4   Cognitive and Emotional Monitoring

Monitoring cognitive and emotional health is a central objective of the proposed framework. This is achieved through naturalistic interaction patterns that incorporate embedded cognitive tasks and continuous emotional assessment, all integrated seamlessly into everyday dialogue.

4.1   Cognitive Stimuli Embedded in Dialogue

To unobtrusively assess cognitive function, the system integrates small assessments into casual conversations, allowing users to be evaluated without the pressure of formal testing. These include:

- Simple memory questions, such as "Can you tell me what you had for breakfast yesterday?" help evaluate short-term memory and attention.
- Prompts like "Can you name five animals starting with 'B'?" assess executive function, vocabulary richness, and mental flexibility. The prompts should be adjusted according to user interest.
- The system narrates a brief story during casual interaction and revisits the story later to assess temporal memory and retention.
- Natural questions like "What day is it today?" or "Where are we right now?" provide insights into disorientation, a common indicator of cognitive decline.

Table 1. Technical Stack Overview

Component	Tools	Description
Language Model	GPT-4 [9], LLaMA2 [10], other transformer-based LLMs	Pre-trained on diverse corpora and fine-tuned on elderly-centric dialogues and medical texts. Optimized for empathy, safety, and interpretability
Speech Modules	Whisper (STT), Google TTS, ElevenLabs [22]	Whisper transcribe diverse speech inputs accurately; Google TTS and ElevenLabs provide expressive voice synthesis for output
Frontend	Flutter, React Native [23]	Cross-platform UI development with support for accessibility features like screen readers, gesture navigation, large fonts, and high contrast
Backend	Python (FastAPI), Node.js [24]	RESTful APIs for efficient communication across components, including logging, failover support, and model version control
Database	MongoDB, PostgreSQL [25]	Stores structured/unstructured data in encrypted format with access control. Compliant with General Data Protection Regulation (GDPR) and Health Insurance Portability and Accountability Act (HIPAA) regulations
Data Sources	Wearables, Social Media, Chat Logs	With consent, enriches user profiles using behavioral and linguistic data to monitor signs of isolation or emotional/cognitive decline

– Queries about daily routines, such as "Did you go for your walk today?" reinforce memory anchoring and allow trend analysis across sessions.

These assessments are inspired by clinically validated instruments such as the MMSE and MoCA [21], but they are reformulated in elder-friendly and culturally relevant language to maintain engagement and avoid user stress.

4.2   Emotional Assessment

Emotional well-being is monitored through multimodal signals derived from voice, language, and behavioral patterns, including:

– Advanced NLP models evaluate the emotional tone of responses, identifying persistent sadness, anxiety, or irritability indicative of depression or emotional stress [26].
– Voice recordings are analyzed for tone, pitch variation, tempo, and energy. Reduced vocal energy or flat intonation may indicate low mood or fatigue [27].

- Repeated phrases or concepts may suggest anxiety, confusion, or cognitive fixation [28]. These patterns are tracked longitudinally.
- The system periodically prompts responses to questions such as "Have you been feeling more tired or down lately?" in a gentle and natural tone to assess mood variability.
- When integrated with smart devices equipped with cameras or microphones, the system may use computer vision to interpret facial expressions, eye contact, and posture to detect signs of emotional withdrawal or stress [29].

Together, these mechanisms support a minute, multi-dimensional understanding of user well-being, allowing for early detection of mental health deterioration.

5 Data Privacy and Ethics

Given the sensitive nature of cognitive and emotional data, the system architecture and policies emphasize privacy, security, and ethical integrity:

- Data Localization and Encryption: Data is stored locally on devices whenever possible. When cloud storage is necessary, information is encrypted end-to-end using GDPR and HIPAA-compliant protocols [30]. Only summarized conversational data and meta-interaction logs are stored locally. High-volume media content is temporarily cached, with user control over what gets retained or purged.
- Federated Learning: The system supports federated learning to enable continuous improvement of LLM performance without transferring raw user data to centralized servers [31]. This preserves privacy while maintaining adaptability. Due to the size of models like GPT-4, federated learning is used only for lightweight model fine-tuning components (e.g., domain-specific layers or memory modules). The base LLM remains centrally updated and accessed through encrypted APIs.
- Informed Consent: Transparent consent workflows ensure users and caregivers understand what data is being collected, how it will be used, and the risks and benefits involved. Users can withdraw consent or delete their data at any time [32].
- Anonymization and Aggregation: When leveraging social media data or additional external sources (with explicit consent), the system applies strict anonymization and aggregation techniques to eliminate identifiable information before use [33]. For instance, if a daughter posts "Mom, we miss you!" on social media, the system anonymizes metadata but relays contextually formatted messages as "Your daughter said: âĂŸMom, we miss you!"' with user consent. Named entity recognition ensures accurate speaker attribution without retaining original identifiers.
- Transparency and Auditability: All data processing activities, model decisions, and system interactions are logged and made accessible through an audit trail [34]. This is crucial for clinician validation, trust-building, and compliance with ethical review standards.

These practices ensure that elderly users can interact with the system confidently, knowing their data is handled responsibly and securely. All data handling procedures adhere to institutional, GDPR, and HIPAA standards.

6   Evaluation Plan

A mixed-method evaluation strategy will be employed to evaluate the effectiveness, usability, and acceptability of the proposed LLM-based framework for cognitive and emotional engagement among elderly individuals. The evaluation will consist of both quantitative and qualitative assessments conducted over a 6-week pilot phase with real-world participants.

6.1   Participants

A pilot study will involve 15âĂŞ20 elderly participants (aged 65 and above) recruited from senior centers and community organizations. While initial testing includes small participants, future iterations will involve larger cohorts to enhance statistical power. Additionally, domain-specific tools such as the mHealth App Usability Questionnaire (MAUQ) and the Health-IT Usability Evaluation Scale (Health-ITUES) will be explored for a more comprehensive usability analysis. This range ensures demographic diversity and provides a more robust foundation for statistical analysis. Participants will be selected to represent a diverse demographic background, including variations in language, digital literacy, and cognitive health status.

6.2   Deployment Setting

Each participant will be provided access to the conversational assistant via a tablet or smartphone, preloaded with the system. Caregivers will be trained to assist and monitor the setup if needed.

Quantitative Evaluation

- System Usability Scale (SUS): A standardized 10-item questionnaire will be administered at the end of the trial to evaluate ease of use, user satisfaction, and overall interface intuitiveness [35].
- MMSE Score Monitoring: Participants' cognitive states will be periodically assessed using MMSE scores [21] derived from embedded interactions. Trends will be evaluated to identify sensitivity and reliability in detecting early signs of cognitive change.
- Sentiment Score Trends: Natural language outputs will be analyzed using pretrained sentiment models to extract weekly emotional states. Metrics such as sentiment polarity and emotional variability will be tracked over time [36].
- Engagement Metrics: Usage frequency, session duration, and interaction type (engagement vs. assessment) will be automatically logged and statistically analyzed.

Qualitative Evaluation. The following criteria can be used for qualitative evaluation of the system:

– Semi-structured interviews will be conducted with participants to understand their perceptions of the system's empathy, usefulness, and comfort level.
– Caregivers will provide insights regarding observed changes in the user's mood, interaction patterns, and trust in the system.
– Qualitative feedback will be transcribed and analyzed to extract recurring themes related to acceptance, perceived benefits, concerns, and suggestions for improvement.

Evaluation Criteria. The following key dimensions will be used to assess the system's success:

– How easily elderly individuals can interact with the system.
– How effectively the system detects and tracks cognitive or emotional changes.
– Emotional connection and trust between the user and the assistant.
– Whether the system respects and reflects the user's language, traditions, and communication norms.

This evaluation plan will inform iterative design improvements, model fine-tuning, and future clinical trials focused on long-term deployment and integration into eldercare infrastructures.

7   Conclusion

The work presents a novel conversational AI framework based on LLMs designed specifically to engage elderly individuals in cognitively, emotionally, and socially enriching dialogues. The proposed system functions both as a virtual companion and a monitoring assistant by identifying subtle cognitive and emotional shifts that may otherwise go unnoticed. By embedding clinically inspired cognitive assessments and emotional analytics into natural conversations, the framework offers a scalable, non-invasive solution for elderly care. The integration of empathetic dialogue, memory personalization, and real-time analytics ensures a more holistic understanding of the elderly well-being of users. Through a multimodal interface and culturally adaptable design, the framework addresses issues of social isolation, accessibility, and mental health awareness among the aging population. In the future, several extensions are planned to enhance the system's usability and clinical integration. As the proposed work is currently a theoretical framework, we plan to develop and implement the complete system in the future. One key direction is the integration of wearable sensor data, which will allow for the inclusion of physiological signals such as heart rate, sleep quality, and mobility patterns to enrich emotional and cognitive monitoring. Additionally, longitudinal clinical studies will be conducted to rigorously validate the system's performance and reliability over extended timeframes. To further personalize interaction, reinforcement learning strategies will be introduced, enabling the system to adapt dynamically to each user's evolving cognitive and emotional state. Lastly, integrating the framework with EHR and telemedicine platforms will ensure seamless collaboration with healthcare professionals, opening the way for broader clinical adoption and real-world impact.

Acknowledgment. The authors acknowledge financial support from the United Arab Emirates University (Grant No. 12R313) through the Big Data Analytics Center, UAEU.

References

1. Chen, M., et al.: A global perspective on risk factors for social isolation in community-dwelling older adults: a systematic review and meta-analysis. Arch. Gerontol. Geriatr. **116**, 105211 (2024)
2. Khan, H.T., Addo, K.M., Findlay, H.: Public health challenges and responses to the growing ageing populations. Public Health Challenges **3**(3), e213 (2024)
3. Albasheer, O., et al.: The impact of social isolation and loneliness on cardiovascular disease risk factors: a systematic review, meta-analysis, and bibliometric investigation. Sci. Rep. **14**(1), 12871 (2024)
4. Aminizadeh, S., et al.: Opportunities and challenges of artificial intelligence and distributed systems to improve the quality of healthcare service. Artif. Intell. Med. **149**, 102779 (2024)
5. Hong, C., et al.: Effect of home-based and remotely supervised combined exercise and cognitive intervention on older adults with mild cognitive impairment (cogito): study protocol for a randomised controlled trial. BMJ Open **14**(8), e081122 (2024)
6. Greenhalgh, T., Wherton, J., Sugarhood, P., Hinder, S., Procter, R., Stones, R.: What matters to older people with assisted living needs? A phenomenological analysis of the use and non-use of telehealth and telecare. Social Sci. Med. **93**, 86–94 (2013)
7. Vaswani, A., et al.: Attention is all you need. In: Advances in Neural Information Processing Systems, vol. 30 (2017)
8. Xygkou, A., et al.: Mindtalker: navigating the complexities of AI-enhanced social engagement for people with early-stage dementia. In: Proceedings of the 2024 CHI Conference on Human Factors in Computing Systems, pp. 1–15 (2024)
9. Sanderson, K.: Gpt-4 is here: what scientists think. Nature **615**(7954), 773 (2023)
10. Touvron, H., et al.: Llama 2: Open foundation and fine-tuned chat models, arXiv preprint arXiv:2307.09288 (2023)
11. Abdollahi, H., Mahoor, M.H., Zandie, R., Siewierski, J., Qualls, S.H.: Artificial emotional intelligence in socially assistive robots for older adults: a pilot study. IEEE Trans. Affect. Comput. **14**(3), 2020–2032 (2022)
12. Miura, C., Chen, S., Saiki, S., Nakamura, M., Yasuda, K.: Assisting personalized healthcare of elderly people: developing a rule-based virtual caregiver system using mobile chatbot. Sensors **22**(10), 3829 (2022)
13. Siricharoen, N.: Creative brain training apps and games can help improve memory, cognitive abilities, and promote good mental health for the elderly, EAI Endorsed Transac (2023)
14. Chen, J., et al.: When large language models meet personalization: perspectives of challenges and opportunities. World Wide Web **27**(4), 42 (2024)

15. Chaturvedi, R., Verma, S., Das, R., Dwivedi, Y.K.: Social companionship with artificial intelligence: recent trends and future avenues. Technol. Forecast. Soc. Chang. **193**, 122634 (2023)

16. Tudor Car, L., et al.: Conversational agents in health care: scoping review and conceptual analysis. J. Med. Internet Res. **22**(8), e17158 (2020). https://doi.org/10.2196/17158

17. Raile, P.: The usefulness of Chatgpt for psychotherapists and patients. Human. Social Sci. Commun. **11**(1), 1–8 (2024)

18. Pandya, A., Lodha, P., Gupta, A.: Technology for early detection and diagnosis of mental disorders: an evidence synthesis. In: Digital Healthcare in Asia and Gulf Region for Healthy Aging and More Inclusive Societies, pp. 37–54. Elsevier (2024). https://doi.org/10.1016/B978-0-443-23637-2.00019-9

19. Almeida, N., Silva, S., Teixeira, A., Ketsmur, M., Guimarães, D., Fonseca, E.: Multimodal interaction for accessible smart homes. In: Proceedings of the 8th International Conference on Software Development and Technologies for Enhancing Accessibility and Fighting Info-exclusion, 2018, pp. 63–70 (2018)

20. Padmaja, P., Ashima Bhatnagar, B.: Designing empathetic interfaces enhancing user experience through emotion: In: Subrata, T., Haipeng, L., Pronaya, B., Samit, B., (eds.) Humanizing Technology With Emotional Intelligence, pp. 47–64. IGI Global (2024). https://doi.org/10.4018/979-8-3693-7011-7.ch004

21. Yu, S., Yu, M.-L., Brown, T., Andrews, H.: Association between older adults functional performance and their scores on the mini mental state examination (mmse) and montreal cognitive assessment (moca). Irish J. Occup. Therapy **46**(1), 4–23 (2018)

22. Bumann, N.: Automated chatbot using speech-to-text and text-to-speech with mobile app integration (2023)

23. Wu, W.: React native vs flutter, cross-platforms mobile application frameworks (2018)

24. Nilsson, E., Demir, D.: Performance comparison of rest vs graphql in different web environments: Node. js and python (2023)

25. Makris, A., Tserpes, K., Spiliopoulos, G., Zissis, D., Anagnostopoulos, D.: Mongodb vs postgresql: a comparative study on performance aspects. GeoInformatica **25**, 243–268 (2021)

26. Kaźmierczak, I., et al.: Natural language sentiment as an indicator of depression and anxiety symptoms: a longitudinal mixed methods study1, Cognition and Emotion, pp. 1–10 (2024)

27. Spreadborough, K.: Emotional tones and emotional texts: a new approach to analyzing the voice in popular vocal song. Music Theory Online **28**(2) (2022). https://doi.org/10.30535/mto.28.2.7

28. Obler, L.K., Albert, M.L.: Language and aging: a neurobehavioral analysis, pp. 107–121. Communication processes and disorders, Aging (1981)

29. Islam, R., Bae, S.W.: Facepsy: an open-source affective mobile sensing system-analyzing facial behavior and head gesture for depression detection in naturalistic settings. In: Proceedings of the ACM on Human-Computer Interaction 8 (MHCI) (2024) 1–32 (2024)

30. Jurczuk, M., Suprunowicz, M.: Consent in data privacy: a general comparison of GDPR and HIPAA. Przegląd Prawniczy Uniwersytetu im. Adam Mickiewicza **16**, 173–194 (2024)

31. Zhang, C., Xie, Y., Bai, H., Yu, B., Li, W., Gao, Y.: A survey on federated learning. Knowl.-Based Syst. **216**, 106775 (2021)

32. Haas, M.A., et al.: ctrlâĂŕ: an online, dynamic consent and participant engagement platform working towards solving the complexities of consent in genomic research. Eur. J. Hum. Genet. **29**(4), 687–698 (2021)
33. Zuo, Z., Watson, M., Budgen, D., Hall, R., Kennelly, C., Al Moubayed, N.: Data anonymization for pervasive health care: systematic literature mapping study. JMIR Med. Inform. **9**(10), e29871 (2021). https://doi.org/10.2196/29871
34. Goktas, P., Grzybowski, A.: Shaping the future of healthcare: ethical clinical challenges and pathways to trustworthy AI. J. Clin. Med. **14**(5), 1605 (2025)
35. Alshamari, M.A., Althobaiti, M.M.: Usability evaluation of wearable smartwatches using customized heuristics and system usability scale score. Future Internet **16**(6), 204 (2024)
36. Mukherjee, P., Badr, Y., Doppalapudi, S., Srinivasan, S.M., Sangwan, R.S., Sharma, R.: Effect of negation in sentences on sentiment analysis and polarity detection. Proc. Comput. Sci. **185**, 370–379 (2021)

Dynamics of Affective States During Takeover Requests in Conditionally Automated Driving Among Older Adults with and Without Cognitive Impairment

Gelareh Hajian[1,2], Ali Abedi[1,2]([✉]), Bing Ye[1,3], Jennifer Campos[1,4], and Alex Mihailidis[1,2,3]

[1] KITE Research Institute, Toronto Rehabilitation Institute,
University Health Network, Toronto, Canada
ali.abedi@uhn.ca

[2] Institute of Biomedical Engineering, University of Toronto, Taranto, Canada

[3] Department of Occupational Science and Occupational Therapy,
University of Toronto, Taranto, Canada

[4] Department of Psychology, University of Toronto, Taranto, Canada

Abstract. Driving is a key component of independence and quality of life for older adults. However, cognitive decline associated with conditions such as mild cognitive impairment (MCI) and dementia can compromise driving safety and often lead to premature driving cessation. Conditionally automated vehicles, which require drivers to take over control when automation reaches its operational limits, offer a potential assistive solution. However, their effectiveness depends on the driver's ability to respond to takeover requests (TORs) in a timely and appropriate manner. Understanding emotional responses during TORs can provide insight into drivers' engagement, stress levels, and readiness to resume control, particularly in cognitively vulnerable populations. This study investigated affective responses, measured via facial expression analysis of valence (emotional tone) and arousal (emotional intensity), during TORs among cognitively healthy older adults and those with cognitive impairment. Facial affect data were analyzed across different road geometries (straight vs. curved) and speeds (50 km/h vs. 100 km/h) to evaluate within- and between-group differences in affective states. Within-group comparisons using the Wilcoxon signed-rank test revealed significant changes in valence and arousal during TORs for both groups. Cognitively healthy individuals showed adaptive increases in arousal under higher-demand conditions, while those with cognitive impairment exhibited reduced arousal and more positive valence in several scenarios. Between-group comparisons using the Mann-Whitney U test indicated that cognitively impaired individuals displayed lower arousal and higher valence than controls across different TOR conditions. These patterns may suggest blunted emotional activation and potentially diminished situational awareness in cognitively impaired drivers. These results also highlight the potential need for adaptive, conditionally automated vehicle systems that can detect affective states and deliver supportive handover strategies tailored to cognitively vulnerable populations.

© The Author(s), under exclusive license to Springer Nature Singapore Pte Ltd. 2025
S. S. Khan et al. (Eds.): IJCAI 2025, CCIS 2620, pp. 29–43, 2025.
https://doi.org/10.1007/978-981-95-0568-5_3

Keywords: Automated Vehicle · Cognitive Impairment · Driver State Monitoring · Older Adults · Human-Machine Interaction · Facial Expression Analysis · Affective Computing

1 Introduction

Driving is a critical enabler of independence, mobility, and social participation among older adults. It is closely associated with quality of life, as it facilitates access to healthcare, social connections, and community engagement [1,2]. However, age-related cognitive decline, particularly due to neurodegenerative conditions such as mild cognitive impairment (MCI) and dementia, can compromise driving safety and often lead to premature driving cessation [3]. This, in turn, has been associated with higher rates of depression and increased three-year mortality [4,5], faster cognitive decline [6], lower quality of life, and disruptions to independence and personal identity [7].

Given the serious consequences of driving cessation for older adults, it is essential to explore transportation alternatives that preserve safe mobility for as long as possible [7,8]. Automated vehicles (AVs) offer a promising solution by reducing the cognitive demands of driving and potentially extending driving ability for individuals with cognitive impairment [7,9–11]. Among these, conditionally automated vehicles (CAVs), classified as Level 3 by the Society of Automotive Engineers [12], enable shared control between the human driver and the automated system. In CAVs, the vehicle manages driving under specific conditions, but the human driver must resume control when a takeover request (TOR) is issued, typically in response to complex or unanticipated scenarios. While CAVs hold promise for prolonging the driving years of older adults, their safety depends on the driver's capacity to re-engage with the driving task effectively during TORs. This challenge is particularly pronounced for older adults with cognitive impairments, who may experience delays in perception, judgment, and motor responses, placing them at elevated risk during these transitions.

Beyond behavioural and physiological indicators of takeover performance, emotional and affective responses may also play a critical role in shaping a driver's readiness and ability to resume control [13–16]. Changes in facial expressions, particularly in terms of valence (the positive or negative emotional tone) and arousal (emotional intensity), can provide insights into a driver's cognitive-affective state during TORs [14–16]. Despite the potential utility of these affective indicators, little is known about how such responses manifest in older adults with and without cognitive impairment during takeover events in automated vehicles.

In this study, facial expressions at the moment of TORs were analyzed across various driving scenarios characterized by differences in road geometry (straight vs. curved) and road speed (50 km/h vs. 100 km/h). Changes in facial affect, specifically valence and arousal, were examined from the period before to during TOR within each group of cognitively healthy older adults and those with cognitive impairment. The aim was to determine whether, and to what extent, changes in affective expression occurred during these safety-critical moments. This within-group analysis was used to reveal how each population emotionally responded to TORs under varying driving conditions, providing a foundation for understanding the potential role of affect in takeover moments. Additionally, a comparison analysis was conducted between the control group (older adults with normal cognition) and those with cognitive impairment during TORs to explore group-level differences in affective responses.

This work represents an initial step in a broader research effort to understand how emotional responses during takeover moments relate to driving performance and physiological markers in older adults, both with and without cognitive impairment. Ultimately, these insights could inform the design of adaptive CAV systems that more effectively support older drivers, particularly those experiencing cognitive decline. The following sections present the literature review, methodology, and results of the facial affect analysis, highlighting key differences within and between groups and across driving conditions during TORs, followed by the conclusion.

2   Literature Review

This section reviews existing literature on analyzing the dynamics of affective states during automated driving, focusing specifically on how emotional factors influence drivers' responses to takeover requests in CAVs.

Du et al. [13] investigated how valence and arousal impact drivers' performance during TORs in Level 3 automated driving scenarios. In a simulated driving environment, 32 healthy university students (young adults, average age 21.4 years, SD = 2.9) were exposed to emotionally charged movie clips designed to induce different emotional states across the valence-arousal spectrum. Participants then experienced unexpected TORs. Takeover time and quality, measured through acceleration, jerk, and time-to-collision, were subsequently assessed. Findings indicated that positive valence significantly improved takeover quality, resulting in smoother and safer transitions (lower acceleration and jerk). However, higher arousal did not improve takeover response times, suggesting that, unlike in manual driving where heightened arousal is often linked to quicker reactions, emotional arousal had less impact in the automated driving context.

Stephenson et al. [14] examined arousal and eye gaze behaviours in 37 older adults (mean age = 68.4 years) during simulated Level 5 autonomous driving using the Wizard of Oz method. Electrodermal activity and heart rate were measured using Empatica E4 wristbands, while eye gaze was tracked using Tobii Pro Glasses 2. Participants experienced both expected and unexpected stops, with the latter simulating safety-critical events. Results indicated increased arousal following unexpected stops. Participants showed longer fixation durations and more frequent fixations toward the central environment (hazard location) during these events, whereas attention shifted more toward the vehicle's human-machine interface during expected stops. This pattern suggests that unexpected events narrow attentional focus and heighten physiological responses. Although the study did not directly examine takeover requests, the unexpected stops served as analogous safety-critical events, likely triggering similar cognitive and affective responses. The findings indicate that heightened arousal and narrowed attention, as observed, are relevant factors influencing takeover performance in TOR scenarios.

Wang et al. [15] analyzed the relationship between driver arousal and takeover performance during automated driving using a high-fidelity driving simulator. A sample of 42 participants across three age groups (young, middle-aged, and older adults) engaged in simulated Level 3 autonomous driving tasks involving TORs under foggy conditions. Arousal was measured using gaze duration and pupil size, recorded via eye-tracking glasses. The results showed that, following TORs, drivers exhibited shorter gaze durations and larger pupil sizes, which were interpreted as indicators of heightened perceptual arousal and improved road information processing.

Du et al. [16] investigated psychophysiological responses to TORs among 102 university students (mean age = 22.9 years) using a high-fidelity driving simulator for SAE Level 3 automated driving. Participants experienced eight TOR scenarios under varying conditions, including different cognitive loads, traffic densities, and TOR lead times. Psychophysiological measures included heart rate, galvanic skin response, gaze behaviours, and facial expressions. Emotional valence and engagement were extracted from facial expressions using the iMotions Affectiva module. The results showed that higher emotional arousal and lower emotional valence (i.e., more negative emotions) were associated with shorter TOR lead times and increased stress. Blink suppression and heart rate acceleration patterns further reflected heightened attention and negative emotional responses in urgent situations. Additionally, higher cognitive load was associated with reduced heart rate variability and narrower gaze dispersion during automated driving, indicating increased mental workload. During takeover, heart rate acceleration occurred more frequently in heavy traffic conditions, suggesting greater attentional demand and stress in more complex environments.

Huang et al. [17] examined how emotional instability affects physiological responses and takeover performance in conditionally automated driving. Forty-two healthy university students (mean age = 24.4 years) completed simulated takeover tasks under both neutral and negative emotional states, which were induced using movie clips. Emotional instability was assessed through subjective personality questionnaires and objective physiological signals, including EEG and ECG, collected via the BIOPAC MP150 system. Valence and arousal were measured using the Self-Assessment Manikin. Participants with higher emotional instability showed greater emotional reactivity, characterized by lower valence, higher arousal, elevated heart rate variability, and increased EEG powerâĂŤindicators of stress, cognitive load, and reduced attention. These participants also demonstrated poorer lateral takeover performance, with higher lateral velocity, acceleration, and turning angles.

Palomares et al. [18] investigated the detection of occupants' emotional states in AVs using physiological signals. Fifty adult drivers (aged 25–55) experienced six Level 4 automated driving scenarios, ranging from smooth driving to critical events such as system failure, within a driving simulator. ECG and EMG signals were collected, and a probabilistic emotional model was applied to estimate valence and arousal in real time. Emotional self-reports using the Self-Assessment Manikin scale confirmed the model's predictions. For instance, the system failure scenario elicited the highest arousal and most negative valence, while the comfort scenario was associated with the lowest arousal and most positive valence.

The reviewed literature demonstrates a growing interest in understanding emotional and physiological responses during conditionally and fully automated driving, with a focus on how affective states influence or reflect takeover performance, trust, and user experience. Prior studies have examined affective dynamics using physiological signals such as heart rate variability, galvanic skin response, EEG, eye tracking, facial expression analysis, and self-reported emotion scales. While these approaches offer valuable insights, most studies have relied on healthy, young adult participants, limiting their relevance to underrepresented, at-risk populations such as older adults and individuals with cognitive impairment. Notably, none of the reviewed work explicitly targets older adults with cognitive impairment, who may face unique cognitive and emotional challenges during high-stakes interactions such as TORs. This is especially important given that emotional expressivity can be diminished in persons with dementia, potentially reducing the reliability of affective cues during takeover scenarios. Additionally, many existing studies rely on wearable sensors or intrusive data collection tools that may be impractical or uncomfortable for long-term, real-world use. In contrast, our work addresses this gap by focusing on older adults with cognitive impairment and analyzing their affective states during TORs using non-intrusive, video-based data. This approach provides a scalable and unobtrusive solution for affect monitoring in automated driving, enabling real-time adaptation of the driving experienceâĂŤfor example, by adjusting the timing or modality of TOR alerts or activating additional support mechanisms based on the driver's emotional and cognitive state. Such personalization is particularly valuable for this population, who may have reduced capacity to cope with abrupt or stressful transitions.

3   Methodology

3.1   Study Setting

This study was conducted using a state-of-the-art driving simulator at the Toronto Rehabilitation Institute, University Health Network, in Toronto, Canada. DriverLab, the most advanced driving simulator in Canada and among the most sophisticated globally, features a full-scale passenger vehicle mounted on a turntable, which can be connected to a six-degree-of-freedom hydraulic motion platform. The vehicle retains its original controls and is equipped with customizable interfaces and integrated measurement tools. A high-resolution projection system displays onto a curved dome surrounding the vehicle, creating a fully immersive and seamless 360-degree field of view, as shown in Fig. 1a.

Figure 1b shows an older adult participant engaged in an automated driving scenario inside the simulator.

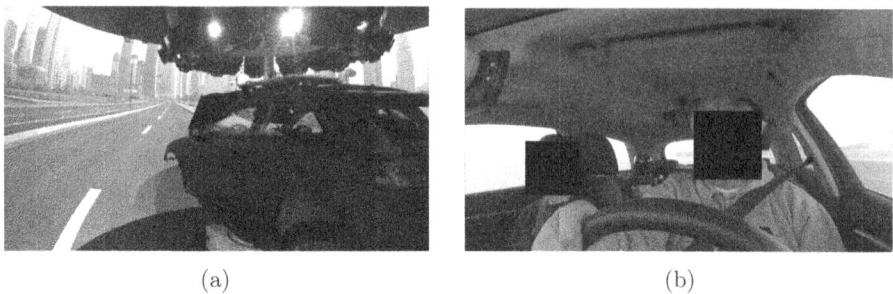

(a)   (b)

Fig. 1. (a) DriverLab simulator at the Toronto Rehabilitation Institute, featuring a full-scale vehicle mounted on a motion platform within a 360-degree immersive projection dome. (b) An older adult participant engaged in an automated driving scenario in DriverLab, with a research assistant seated in the passenger seat.

Study Design. The study followed a 2 (cognitive status) $\times$ 2 (road geometry) $\times$ 2 (speed limit) $\times$ 2 (lighting condition) factorial design. The between-subjects factor was cognitive status, consisting of cognitively healthy individuals (control, $n = 18$) and individuals with cognitive impairment, including those with MCI and very mild dementia ($n = 5$). The within-subjects factors were road geometry (straight vs. curved) and speed limit (50 km/h vs. 100 km/h). Although both daytime and nighttime scenarios were included in the experimental design, only data from daytime conditions were analyzed here due to reduced visibility in nighttime scenarios.

Study Participants. To be included in the healthy control group, participants were required to be 65 years or older, have a valid driver's license, self-report normal motor and sensory functioning, and score $\geq$ 26 on the Montreal Cognitive Assessment (MoCA) [19]. For inclusion in the cognitively impaired group, participants with MCI or very mild dementia had to be 65 years or older, self-report a clinical diagnosis of MCI or very mild dementia from a healthcare provider, report no motor or sensory impairments, score 0.5 on the Clinical Dementia Rating Scale (CDR) [20], and either hold a valid driver's license or have voluntarily stopped driving within the past 18 months. One participant in the cognitively impaired group did not complete the CDR assessment due to the absence of an available informant, which is required for CDR administration. In this case, group assignment was based on the participant's clinical diagnosis provided by their physician.

Study Procedures. The study involved two visits to DriverLab. The first visit included a simulator adaptation session and the administration of screening and baseline assessments. Screening procedures involved obtaining informed consent and completing the Health History Questionnaire, which covered age, medication use, and general health status. Baseline assessments evaluated cognitive and functional abilities across key domains, including attention, memory, executive function, and processing speed. During this visit, participants also completed a 10-minute adaptation drive to familiarize themselves with the simulator environment and Level 3 automation, while reducing the likelihood of simulator sickness. This session included a practice takeover task to ensure participants were comfortable and capable of responding to takeover requests. Participants were allowed to repeat the practice session if needed. Individuals who experienced significant simulator sickness during this visit did not proceed to the second visit.

The second visit included two 12-minute conditionally automated driving sessions under varying lighting (daytime or nighttime) and road conditions. In each session, the vehicle was initially operated by the automated system. At predetermined points, an audio-visual takeover request (TOR) was issued, prompting participants to resume manual control. Each session included four TOR events, covering different combinations of road geometry (straight and curved) and speed limits (50 km/h and 100 km/h). One session was conducted during the daytime and the other at nighttime; session order was counterbalanced across participants. Breaks were provided between scenarios, during which participants were screened for simulator sickness using the Fast Motion Sickness Scale (FMSS) [21]. Scheduled rest periods supported recovery and helped minimize simulator sickness. If mild symptoms were reported, additional breaks were offered; sessions were discontinued in cases of severe symptoms. To enhance comfort and reduce the likelihood of dropout, the environment was kept cool with steady airflow. For the purpose of this paper, only data from the daytime sessions were analyzed. Nighttime facial expression data were excluded due to unreliable visibility conditions.

The study was approved by the University Health Network Research Ethics Board (REB #20-5090). Informed consent was obtained from all participants using plain-language forms. For individuals with MCI or mild dementia, a caregiver or informant co-signed the consent form as an additional safeguard. Participants were monitored throughout the study, and sessions were paused or discontinued if necessary to ensure their well-being.

3.2   Analysis of Affective State Dynamics

Video recordings of drivers during conditionally automated driving sessions were captured using a GoPro camera at a resolution of 1920 × 1080 pixels and 48 frames per second. The videos were downsampled to 16 frames per second using FFmpeg [22] to reduce computational load while maintaining sufficient temporal resolution for affective analysis. This frame rate was selected based on prior research indicating that 10–15 frames per second is typically adequate for capturing facial expression dynamics [23]. The videos were then spatially cropped using FFmpeg [22] to exclude the front-seat passenger (the research assistant) and retain only the driver's region in each frame.

Facial region detection was performed using MediaPipe Face Detection [24], which generated bounding boxes around the face in each frame. These cropped face images were processed using EmoFan (Emotion Face Alignment Network) [25], a deep learning model pre-trained on AffectNet [26], to compute frame-level continuous values of valence and arousal. Sixteen valence and sixteen arousal scores were extracted per second of video, with each score represented as a real-valued number ranging from −1 to +1, enabling high-resolution temporal tracking of affective states throughout the driving session.

Valence represents the positivity or negativity of an emotional state, indicating its pleasantness or unpleasantness, while arousal reflects the intensity of the emotion, ranging from calm or deactivated to excited or activated states. Together, these two dimensions form the basis of the circumplex model of affect [27], a widely used framework in affective computing (see Fig. 2).

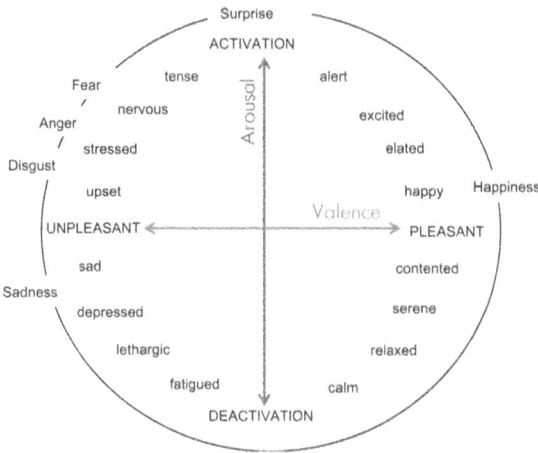

Fig. 2. Russell's circumplex model of affect [27], with valence on the horizontal axis and arousal on the vertical axis.

3.3   Statistical Analysis

Within-Group Analysis. To examine the affective dynamics of drivers during TORs, the primary dependent variables, valence and arousal (derived from facial expression data), were compared between the periods before and during TOR events. The Wilcoxon signed-rank test with a two-sided hypothesis was used to assess whether affective responses during TORs differed significantly from those observed during the preceding automated driving period. This non-parametric test was chosen due to the non-normal distribution and ordinal nature of the data, making it suitable for comparing two related samples. Analyses were conducted separately for cognitively healthy older adults and those with cognitive impairment. Both p-values and standardized mean differences (SMDs) were reported to evaluate the statistical and practical significance of the observed changes.

Between-Group Analysis. To compare affective responses between independent groups, specifically, healthy older adults and those with cognitive impairment after TORs, the Mann-Whitney U test was applied using a two-sided hypothesis. This test was appropriate due to the non-normal distribution and unequal group sizes (control group: $n = 18$; cognitive impairment group: $n = 5$). The analysis evaluated whether TORs were associated with statistically significant changes in emotional responses across groups.

4   Results and Discussion

Facial expressions during TORs were analyzed to assess valence and arousal in older adults with and without cognitive impairment. Changes from pre-TOR to TOR were examined within each group across road geometries (straight vs. curved) and speed (low speed (50 km/h) vs. high speed (100 km/h)). The aim was to evaluate within-group affective changes and between-group differences. An example of changes in valence and arousal values for an older adult pre- and during a TOR is shown in Fig. 3.

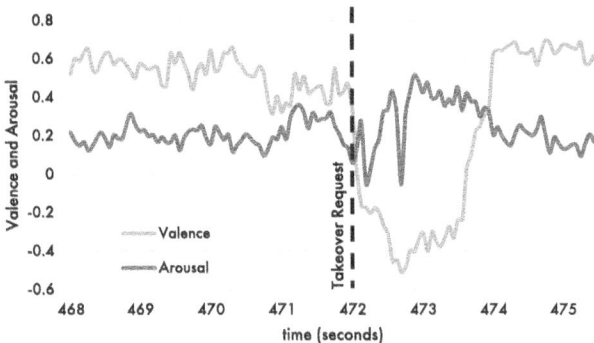

Fig. 3. Changes in valence and arousal values for an older adult with MCI before and after a takeover request (TOR) occurring at second 472, during the high-speed (100km/h) straight condition.

4.1 Facial Affect Analysis Results

Statistical Analysis Results - Within-Group Analysis. Tables 1 (a) and
(b) present within-group differences in valence and arousal before and during TORs across four road conditions: low speed (50 km/h)-straight, low speed
(50 km/h)-curved, high speed (100 km/h)-straight, and high speed (100 km/h)-
curved, for cognitively healthy older adults and those with cognitive impairment.
For each condition, p-values from the Wilcoxon signed-rank test and SMDs are
reported. A positive SMD indicates an increase in valence or arousal from the
automated driving period to the TOR moment, whereas a negative SMD indicates a decrease.

Table 1. Changes in (a) valence and (b) arousal during takeover requests (TORs),
compared to time periods preceding TORs, across different TOR conditions in healthy
older adults and those with cognitive impairment. p-values from the Wilcoxon signed-
rank test and standardized mean differences (SMDs) are reported. The expected cognitive demands for the different TOR conditions were considered as follows: low for low
speed-straight, medium for both low speed-curved and high speed-straight, and high
for high speed-curved.

(a) Valence

TOR condition	Cognitively Healthy				Cognitively Impaired			
	Before	During	p-value	SMD	Before	During	p-value	SMD
Low speed-straight	-0.22 ± 0.11	-0.16 ± 0.13	< 0.001	0.43	-0.11 ± 0.11	-0.14 ± 0.10	<0.001	-0.29
Low speed-curved	-0.23 ± 0.10	-0.18 ± 0.12	<0.001	0.33	-0.13 ± 0.10	-0.04 ± 0.20	<0.001	0.57
High speed-straight	-0.21 ± 0.11	-0.26 ± 0.16	<0.001	-0.35	-0.08 ± 0.27	-0.07 ± 0.32	<0.001	0.15
High speed-curved	-0.24 ± 0.20	-0.22 ± 0.24	<0.001	-0.06	-0.10 ± 0.12	-0.24 ± 0.07	<0.001	-1.11

(b) Arousal

TOR condition	Cognitively Healthy				Cognitively Impaired			
	Before	During	p-value	SMD	Before	During	p-value	SMD
Low speed-straight	0.16 ± 0.11	0.12 ± 0.12	<0.001	-0.30	0.15 ± 0.07	0.12 ± 0.09	<0.001	-0.35
Low speed-curved	0.14 ± 0.11	0.17 ± 0.12	0.0011	0.25	0.16 ± 0.07	0.11 ± 0.09	<0.001	-0.52
High speed-straight	0.15 ± 0.11	0.17 ± 0.15	<0.001	0.15	0.14 ± 0.07	0.16 ± 0.14	0.0012	0.20
High speed-curved	0.15 ± 0.12	0.17 ± 0.13	<0.001	0.20	0.14 ± 0.07	0.14 ± 0.12	0.0224	0.01

Valence– Significant changes in valence were observed in all TOR conditions
for both groups ($p < 0.001$ for all comparisons). Among cognitively healthy
older adults, the greatest increases in valence occurred in the low speed-straight
(SMD = 0.43) and low speed-curved (SMD = 0.33) conditions, indicating moderate positive shifts in emotional tone. Conversely, valence decreased slightly in
high speed conditions, with a small negative shift in high speed-straight (SMD =
-0.35) and a minimal change in high speed-curved (SMD = -0.06), suggesting
mildly heightened unpleasant emotion or reduced emotional positivity during
high-speed TORs. For the cognitively impaired group, valence also changed significantly across all conditions, but the direction and magnitude of effects varied

more substantially. A notable increase was observed in the low speed-curved condition (SMD = 0.57), suggesting a strong shift toward more positive emotional tone. This may reflect greater comfort in moderate-demand scenarios or a different perception of the event's complexityâĂŤpotentially indicating that participants with cognitive impairment did not perceive the event as being as risky or demanding as it objectively was. In contrast, a sharp decline in valence was found in the high speed-curved condition (SMD = −1.11), indicating a marked increase in unpleasant emotional experience during the most cognitively demanding TORs. Smaller shifts were seen in low speed-straight (SMD = −0.29) and high speed-straight (SMD = 0.15), suggesting minimal or inconsistent emotional responses in those conditions.

Arousal– Arousal responses also showed statistically significant changes in most conditions, for both groups. Among cognitively healthy older adults, arousal decreased in the low speed-straight condition (SMD = −0.30), indicating a reduction in emotional activation. This may suggest that the event was perceived as low risk, leading to a more relaxed or disengaged state. In contrast, arousal increased in all other conditions: low speed-curved (SMD = 0.25), high speed-straight (SMD = 0.15), and high speed-curved (SMD = 0.20). These increases suggest greater attentional engagement and heightened alertness in response to more cognitively demanding scenarios, particularly those involving curves or high speeds. This pattern reflects an adaptive emotional response, with participants modulating arousal in accordance with perceived task complexity and urgency. In contrast, older adults with cognitive impairment demonstrated more variable arousal responses. Arousal significantly decreased in both low speed-straight (SMD = −0.35) and low speed-curved (SMD = −0.52) conditions, with the larger decline in the curved scenario suggesting blunted emotional engagement even in moderately demanding contexts. Arousal increased slightly in the high speed-straight condition (SMD = 0.20), indicating some degree of cognitive or emotional engagement. However, arousal remained almost unchanged in the high speed-curved condition (SMD = 0.01), despite it being the most cognitively demanding scenario. This minimal response could indicate impaired situational awareness or a failure to perceive the complexity of the driving task.

Overall, cognitively healthy participants exhibited more consistent and adaptive affective responses, characterized by increased arousal and moderately increased valence in low speed-curved conditions, along with small arousal increases in high-speed scenarios. In contrast, individuals with cognitive impairment showed greater variability in both valence and arousal. While valence increased substantially in the low speed-curved condition, it sharply decreased under high speed-curved scenarios. Arousal responses were generally blunted across conditions in the cognitively impaired group, with significant decreases even in moderately demanding contexts and minimal change in the most complex condition. These patterns suggest reduced emotional engagement or less effective modulation of affective states during TORs, particularly under higher cognitive demand.

Statistical Analysis Results - Between-group Analysis Table 2 presents the results of the between-group comparison of valence and arousal values following TORs across different driving conditions. Comparisons were made between those with cognitive impairment and cognitively healthy older adults using the Mann-Whitney U test. For each condition, p-values and SMDs are reported.

Table 2. Differences in valence and arousal values between older adults with cognitive impairment and cognitively healthy older adults in time periods during takeover requests (TORs) across different TOR conditions. Reported values include p-values from the Mann-Whitney U test and standardized mean differences (SMDs).

	Valence		Arousal	
TOR condition	p-value	SMD	p-value	SMD
Low speed-straight	<0.001	0.36	0.0946	0.13
Low speed-curved	<0.001	0.91	<0.001	−0.42
High speed-straight	<0.001	0.86	<0.001	−0.20
High speed-curved	<0.001	−0.07	0.0029	−0.22

Valence– Significant between-group differences in valence were found across all TOR conditions ($p < 0.001$). The largest differences were observed in the low speed-curved (SMD = 0.91) and high speed-straight (SMD = 0.86) conditions, where the cognitively impaired group showed higher valence values than the control group. This suggests that, in these scenarios, the cognitively impaired participants exhibited a more positive or less negative emotional tone following TORs. This may reflect less perceived threat or stress to moderately demanding events (curved and high speed cases) in cognitively impaired participants than in controls. Moderate group differences were also observed in the low speed-straight condition (SMD = 0.36), while the smallest and only slightly negative difference occurred in the high speed-curved condition (SMD = -0.07), indicating relatively similar valence levels between the groups.

Arousal– Arousal differences between groups were generally smaller in magnitude but statistically significant in most conditions. In low speed-curved (SMD = −0.42), high speed-straight (SMD = −0.20), and high speed-curved (SMD = −0.22) conditions, the cognitively impaired group exhibited lower arousal levels compared to the control group, suggesting reduced emotional activation, alertness, or attentional response in this group in conditions requiring moderate to high engagement. In the low speed-straight condition, the difference was not statistically significant ($p = 0.0946$, SMD = 0.13), indicating comparable levels of arousal across groups in simpler road scenarios.

These results highlight important affective differences during TORs between cognitively healthy and impaired older adults. While the cognitively impaired group often experienced more positive valence, they also showed reduced arousal in moderately and highly demanding conditions, suggesting a possible mismatch between emotional tone and task complexity. This pattern may reflect impaired risk perception, reduced emotional reactivity, or altered engagement.

4.2   Limitation

A key limitation of this study is the small sample size for the cognitively impaired group, which may reduce statistical power, increase the risk of Type I and II errors, lead to unstable effect size estimates, and limit the generalizability of between-group comparisons. Future work should aim for larger, balanced groups (at least 15–20 participants per group) to improve reliability and generalizability.

Additionally, this study relied on facial expression data as the sole indicator of affective states. While facial expressions provide valuable insight into emotional responses, they may not fully capture the complexity of internal emotional or cognitive states, particularly in older adults with cognitive impairment, who may exhibit blunted or atypical facial expressions. Future studies should include larger and more diverse samples and consider integrating multimodal physiological measures (e.g., heart rate variability via photoplethysmography and gaze patterns via eye-tracking) to gain a more comprehensive understanding of emotional and cognitive readiness during TORs.

Moreover, facial affect estimation models are often trained on younger, neurotypical populations and may not generalize well to older adults, particularly those with neurodegeneration, whose facial expressions can be subtler, atypical, or less dynamically expressed. This introduces potential biases in affect recognition.

Finally, although data collection took place in a highly immersive driving simulator, the setting cannot fully replicate the complexities of real-world driving. Factors such as lighting variability, driver fatigue, and unexpected external distractions were not entirely represented. These real-world elements may influence both emotional responses and the performance of models trained under controlled conditions. Future work should aim to validate the robustness of these models in more naturalistic environments to improve generalizability.

5   Conclusion

This study examined affective responses, specifically valence and arousal, during TORs in CAVs among cognitively healthy older adults and those with cognitive impairment. Within-group analyses revealed that both groups exhibited significant emotional shifts during TORs, although the direction and intensity of these changes varied depending on road context. Notably, cognitively impaired participants showed increased valence in two conditions, potentially reflecting a diminished perception of risk, and lower arousal in low-speed settings, suggesting reduced emotional engagement or alertness during these safety-critical transitions. Between-group comparisons further supported these patterns, with the cognitively impaired group displaying a mismatch between emotional tone and intensity relative to their healthy counterparts.

These findings highlight important differences in how older adults with cognitive impairment emotionally experience TORs, raising concerns about their readiness to resume control in CAVs. The observed blunting of arousal, particularly in more demanding scenarios, may compromise takeover performance and highlights the need for adaptive vehicle systems that can better support this population. Future work will examine the relationship between facial expressions and takeover performance and explore how affective and physiological signals can be integrated to predict takeover readiness.

Acknowledgment. This research was supported by the Canadian Institutes of Health Research (CIHR), AGE-WELL (Canada's Technology and Aging Network), and the Alzheimer's Society of Canada.

References

1. Maresova, P., et al.: Challenges and opportunity in mobility among older adults-key determinant identification. BMC Geriatr. **23**(1), 447 (2023)
2. Qin, W., Xiang, X., Taylor, H.: Driving cessation and social isolation in older adults. J. Aging Health **32**(9), 962–971 (2020)
3. Teasdale, N., et al.: Older adults with mild cognitive impairments show less driving errors after a multiple sessions simulator training program but do not exhibit long term retention. Front. Human Neurosci. **10** (2016). https://doi.org/10.3389/fnhum.2016.00653
4. Fonda, S.J., Wallace, R.B., Herzog, A.R.: Changes in driving patterns and worsening depressive symptoms among older adults. J. Gerontol. Series B **56**(6), S343–S351 (2001)
5. Chihuri, S., et al.: Driving cessation and health outcomes in older adults. J. Am. Geriatr. Soc. **64**(2), 332–341 (2016)
6. Choi, M., Lohman, M.C., Mezuk, B.: Trajectories of cognitive decline by driving mobility: Evidence from the health and retirement study. Int. J. Geriatr. Psychiatry **29**(5), 447–453 (2014)
7. Sanford, S., Naglie, G., Cameron, D.H., Rapoport, M.J.: Subjective experiences of driving cessation and dementia: a meta-synthesis of qualitative literature. Clin. Gerontol. **43**(2), 135–154 (2020). https://doi.org/10.1080/07317115.2018.1483992
8. Chee, J.N., et al.: Update on the risk of motor vehicle collision or driving impairment with dementia: a collaborative international systematic review and meta-analysis. Am. J. Geriatric Psych. **25**(12), 1376–1390 (2017). https://doi.org/10.1016/j.jagp.2017.05.007
9. Shergold, I., Wilson, M., Parkhurst, G.: The mobility of older people and the future role of connected autonomous vehicles. J. Transp. Health (2016), retrieved from https://uwe-repository.worktribe.com
10. Jang, R.W., et al.: Family physicians attitudes and practices regarding assessments of medical fitness to drive in older persons. J. Gen. Intern. Med. **22**(4), 531–543 (2007)
11. Hajian, G., Ye, B., Haghzare, S., Campos, J.L., Mihailidis, A.: Conditionally automated vehicles and cognitive challenges: assessing the safety of conditionally automated vehicles for older adults with cognitive challenges. Alzheimer's & Dementia **20**, e091988 (2024)

12. SAE International, "Taxonomy and definitions for terms related to driving automation systems for on-road motor vehicles (sae standard j3016)," 2018. https://www.sae.org/standards/content/j3016_201806/
13. Du, N., et al.: Examining the effects of emotional valence and arousal on takeover performance in conditionally automated driving. Transport. Res. Part C: Emerg. Technol. **112**, 78–87 (2020)
14. Stephenson, A.C., et al.: Effects of an unexpected and expected event on older adults autonomic arousal and eye fixations during autonomous driving. Front. Psychol. **11**, 571961 (2020)
15. Wang, Q., Chen, H., Gong, J., Zhao, X., Li, Z.: Studying driver s perception arousal and takeover performance in autonomous driving. Sustainability **15**(1), 445 (2023)
16. Du, N., Yang, X.J., Zhou, F.: Psychophysiological responses to takeover requests in conditionally automated driving. Accident Anal. Prevent. **148**, 105804 (2020)
17. Huang, J., Long, X., Qi, C., Hu, L., Gao, K.: Differences in driver takeover performance and physiological responses in conditionally automated driving: Links to emotional instability. Transport. Res. F: Traffic Psychol. Behav. **105**, 73–86 (2024)
18. Palomares, N., et al.: Detection of occupant emotion in automated vehicles under different driving conditions. Transport. Res. Proc. **72**, 3917–3924 (2023)
19. Nasreddine, Z.S., et al.: The Montreal cognitive assessment, MOCA: a brief screening tool for mild cognitive impairment. J. Am. Geriatr. Soc. **53**(4), 695–699 (2005)
20. Morris, J.C.: The clinical dementia rating (CDR) current version and scoring rules. Neurology **43**(11), 2412–2412 (1993)
21. Keshavarz, B., Hecht, H.: Validating an efficient method to quantify motion sickness. Hum. Factors **53**(4), 415–426 (2011)
22. FFmpeg Developers, "Ffmpeg," https://ffmpeg.org/, 2023, version 5.1.2 or later
23. Sariyanidi, E., Gunes, H., Cavallaro, A.: Automatic analysis of facial affect: a survey of registration, representation, and recognition. IEEE Trans. Pattern Anal. Mach. Intell. **37**(6), 1113–1133 (2014)
24. Lugaresi, C.: Mediapipe: A framework for building perception pipelines. In: Proceedings of the IEEE/CVF Conference on Computer Vision and Pattern Recognition Workshops, 2019, pp. 2278–2282 (2019)
25. Toisoul, A., Kossaifi, J., Bulat, A., Tzimiropoulos, G., Pantic, M.: Estimation of continuous valence and arousal levels from faces in naturalistic conditions. Nature Mach. Intell. **3**(1), 42–50 (2021)
26. Mollahosseini, A., Hasani, B., Mahoor, M.H.: Affectnet: A database for facial expression, valence, and arousal computing in the wild. IEEE Trans. Affect. Comput. **10**(1), 18–31 (2017)
27. Russell, J.A.: A circumplex model of affect. J. Pers. Soc. Psychol. **39**(6), 1161 (1980)

Continuous Monitoring of Emotional Decline in Older Chinese Adults via Hierarchical Temporal Inference

Hongru Ma[1(✉)], Sihang Zhang[2], and Yanjie Liang[3]

[1] School of Software, Beihang University, Beijing, China
mahongru@buaa.edu.cn
[2] Dashcoding Research, Shenzhen, China
[3] School of Software, Shandong University, Jinan, China

Abstract. Emotional decline is a gradual and often overlooked process among older adults, with serious consequences for mental health and quality of life. Existing emotion detection methods typically rely on physiological sensors or isolated speech recordings. However, these approaches lack temporal continuity and are unsuitable for long-term monitoring. To address these limitations, we propose CEDME (Continuous Emotion Decline Monitoring for the Elderly), a multimodal and temporally structured framework that continuously monitors emotional well-being from naturalistic daily speech. CEDME captures fine-grained vocal and lexical signals through gated fusion, and models emotional trajectories over time via hierarchical temporal inference across both intra-day and cross-day scales. To support this effort, we construct a Chinese elderly speech dataset (CESD) comprising 14 consecutive days of conversational recordings from older adults in China. Extensive experiments on long-range emotional monitoring show that CEDME consistently outperforms strong baselines, confirming its effectiveness as a scalable, non-intrusive, and clinically interpretable solution for real-world monitoring of emotional decline in the elderly.

Keywords: Emotion monitoring in older adults · Multimodal fusion · Hierarchical temporal inference

1 Introduction

As China rapidly enters an aging society [37], emotional decline, which refers to the progressive deterioration of emotional expressiveness, regulation, and responsiveness, has become a silent yet widespread crisis among older adults, often co-occurring with cognitive deterioration, depressive symptoms, and social withdrawal [36]. A growing number of elderly individuals now live alone or have minimal daily interaction, especially in urban-rural fringe areas and depopulating rural regions, leading to chronic loneliness, emotional blunting, and heightened

© The Author(s), under exclusive license to Springer Nature Singapore Pte Ltd. 2025
S. S. Khan et al. (Eds.): IJCAI 2025, CCIS 2620, pp. 44–59, 2025.
https://doi.org/10.1007/978-981-95-0568-5_4

risks of mental health deterioration. These conditions typically progress insidiously and go unnoticed until reaching critical thresholds, when interventions are less effective or too late. Without timely detection and continuous monitoring, emotional decline can escalate into severe affective disorders, accelerate cognitive deterioration, and drastically reduce quality of life.

To support emotion monitoring in older adults, researchers have developed psychophysiological methods that infer internal affective states based on biological signals. Common modalities include electroencephalography [8,15], electrodermal activity [25,32], and heart rate variability [19], which reflect neural, skin, and cardiac signals, respectively. Other approaches, such as functional near-infrared spectroscopy [13], facial analysis [20], and gaze tracking [35], also infer emotions from brain and behavioral cues. However, these techniques face critical limitations in real-world elderly care, as they often produce noisy signals, are sensitive to movement, consistently struggle to detect subtle emotional changes, and typically require controlled environments [12]. Moreover, older adults, especially those living alone in geographically dispersed households, often resist intrusive devices [3], and even when willing, their scattered living arrangements make centralized deployment extremely difficult.

To overcome theses limitations, recent studies [21,24,29] have increasingly investigated speech-based, non-intrusive approaches for emotion recognition in the elderly. These methods typically analyze brief voice recordings collected through daily interviews, scripted tasks, or spontaneous interactions, and apply classification or regression models to infer emotional states [30]. However, most existing methods operate on isolated utterances or single-session recordings, without modeling how emotional states evolve over time. In practice, emotional decline in older adults is rarely abrupt, as it emerges gradually and is reflected in subtle shifts in vocal tone, fluency, or lexical patterns across multiple days, which recent studies fail to capture due to their snapshot-based design.

Across existing approaches, the core limitation remains unchanged. Emotional states in elderly individuals are often treated as isolated observations, without accounting for how subtle affective cues accumulate within a day and evolve across days, leading to a fundamental blind spot in capturing the gradual and progressive nature of emotional decline in older adults. As a result, existing methods focus narrowly on discrete detection rather than continuous monitoring and fail to meet the long-term demands of elderly care.

To address the limitations of existing approaches, in this paper, we propose CEDME (Continuous Emotion Decline Monitoring for the Elderly), a temporally structured and multimodally grounded framework for continuous monitoring of emotional decline in elderly individuals. To the best of our knowledge, this is the first framework that enables temporally structured and continuous monitoring of emotional trajectories from naturalistic daily speech in real-world aging scenarios. CEDME captures emotional trajectories through hierarchical temporal modeling across both intra-day and inter-day scales, and integrates prosodic and linguistic features via multimodal fusion. The resulting well-being scores inform care recommendations, offering a scalable and non-intrusive solution for elderly mental health support. Extensive experiments demonstrate the effectiveness of CEDME in capturing longitudinal emotional dynamics and providing reliable well-being assessments in real-world elderly monitoring scenarios.

Accordingly, this paper makes the following major contributions: (1) We propose CEDME, the first framework for continuous monitoring of emotional decline in elderly individuals based on daily speech. (2) We design a hierarchical temporal modeling strategy that captures fine-grained emotional dynamics both within and across days, enabling continuous tracking of emotional changes over time. (3) We conduct extensive experiments on real-world elderly speech data, demonstrating that CEDME outperforms existing baselines in longitudinal emotion monitoring and well-being assessment.

2   Related Work

Intrusive emotion recognition methods for older adults typically utilize physiological signals collected via wearable or contact-based sensors. Prior research has explored multimodal strategies combining electroencephalography (EEG), electrodermal activity (EDA), and heart rate variability (HRV) data, employing architectures such as recursive mapping and vision transformers for improved accuracy [8,11,15,20]. Systems integrating facial expressions, voice, gaze, and cardiovascular cues within virtual coaching platforms have also shown promise for real-time affective assessment [25]. However, the dependency on specialized hardware, participant adherence, and limited ecological validity hampers their practical deployment, while their intrusive nature raises ethical concerns in long-term care scenarios.

By contrast, non-intrusive methods have shifted attention to speech-based analysis drawn from brief interviews or spontaneous narratives. Recent studies employ bimodal frameworks to classify arousal and valence using low-level acoustic and linguistic features [30]. Others estimate subjective well-being from natural speech, using machine learning models trained on WHOQOL-derived labels and prosodic-spectral descriptors [9]. LSTM-based fusion models and short-interview embeddings have also been proposed for one-time assessments [2,16,24]. While these methods offer greater accessibility and user acceptance, they generally lack temporal continuity and are limited in tracking long-term emotional dynamics essential for early intervention.

3   Methods

3.1   Problem Formulation

We address the task of Chinese elderly emotional decline monitoring under a low-resource, multimodal setting. The input $\mathcal{I}$ consists of conversational data collected over T days. Each day includes S_t short speech segments, where each segment comprises a raw audio recording and its corresponding transcribed text.

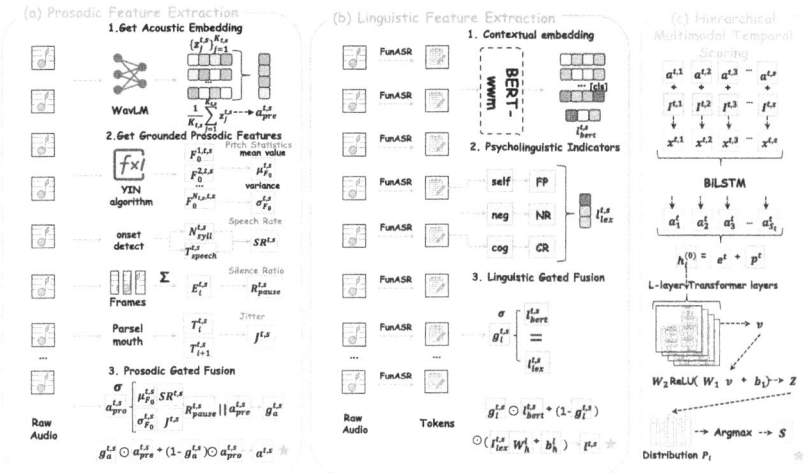

Fig. 1. Overview of CEDME, which comprises three core modules.

These segments are obtained from daily interactions and the goal is to predict an aggregated emotional well-being score s from the entire sequence:

$$f(\mathcal{I}) \rightarrow s \in \{1, 2, \ldots, 10\}, \tag{1}$$

where higher values of s indicate more stable or positive emotional states.

3.2 Data Collection and Pre-Processing

We collect T-day conversational data from elderly individuals living alone or in community care. Each participant engages in brief daily dialogues captured through four channels: routine phone calls, smart speaker interactions, voice messages, and in-person conversations, recorded via portable devices. Let RA_t^s denote the raw audio of the s-th segment on day t. We employ FunASR[1], a Mandarin-optimized automatic speech recognition (ASR) framework, to transcribe each segment and obtain its token sequence:

$$\text{FunASR}(RA_t^s) \rightarrow \mathbf{w}_t^s = \{w_1^{t,s}, w_2^{t,s}, \ldots, w_{M_{t,s}}^{t,s}\}, \tag{2}$$

where $\mathbf{w}_t^s$ denotes the token sequence for segment s on day t. In Chinese ASR transcripts, each token $w_i^{t,s}$ is the smallest lexical unit produced by the tokenizer, which can be either a whole word or a sub-word fragment depending on the segmentation rule. The daily multimodal input comprises S_t paired streams $\{(RA_t^s, \mathbf{w}_t^s)\}_{s=1}^{S_t}$, and the full sequence is denoted as $\mathcal{I} = \{\{(RA_t^s, \mathbf{w}_t^s)\}_{s=1}^{S_t}\}_{t=1}^{T}$. For acoustic processing, each segment RA_t^s is split into overlapping frames using

---

[1] https://github.com/modelscope/FunASR?utm_source=chatgpt.com.

a 25 ms window with 10 ms shift at 16 kHz, resulting in $L = 400$ samples per frame. Each frame is denoted by:

$$\mathbf{r}_i^{t,s} = \{r_{i,1}^{t,s}, r_{i,2}^{t,s}, \ldots, r_{i,L}^{t,s}\}, \quad \mathbf{r}_i^{t,s} \in \mathbb{R}^L. \tag{3}$$

3.3 Overview

As shown in Fig. 1, CEDME comprises three core modules to support multimodal long-term emotional decline monitoring: (a) *Prosodic Feature Extraction Module* encodes acoustic cues related to vocal control and affective state; (b) *Linguistic Feature Extraction Module* captures contextual semantics and psychologically salient lexical patterns; and (c) *Hierarchical Multimodal Temporal Scoring Module* performs multimodal fusion and hierarchical temporal reasoning within and across days, ultimately producing an emotional well-being score.

3.4 Prosodic Feature Extraction Module

Prosodic features in elderly speech, including reduced pitch variability, slower speech rate, and prolonged pauses, convey important information about vocal control, cognitive fluency, and affective state [26, 29, 33]. These non-verbal cues are less influenced by linguistic habits and serve as a useful complement to lexical features, particularly in spontaneous and low-resource settings [2]. To capture these cues effectively, CEDME combines two sources of prosodic information: (1) pretrained acoustic embeddings that encode high-level vocal dynamics, and (2) physiologically grounded prosodic features that quantify vocal signals such as pitch variability and pause behavior. This hybrid representation enhances CEDME's ability to capture both high-level prosodic patterns and task-relevant acoustic cues. We describe both components in the following subsections.

Pretrained Acoustic Embeddings. To capture high-level prosodic and articulatory patterns without manual feature design, we extract latent acoustic representations from a self-supervised pretrained model. Specifically, we employ WavLM-Large[2], a self-supervised model pretrained for multilingual speech representation learning, to encode each audio segment into a sequence of frame-level hidden states. The pretrained model is applied to the entire waveform RA_t^s, and an utterance-level acoustic embedding is then obtained by mean pooling over all hidden states:

$$\{\mathbf{z}_j^{t,s}\}_{j=1}^{K_{t,s}} = \text{WavLM}(RA_t^s), \quad \mathbf{a}_{\text{pre}}^{t,s} = \frac{1}{K_{t,s}} \sum_{j=1}^{K_{t,s}} \mathbf{z}_j^{t,s} \in \mathbb{R}^d, \tag{4}$$

where $\mathbf{z}_j^{t,s}$ denotes the hidden state of the j-th output frame for segment s on day t, and $K_{t,s}$ is the total number of frames. The embedding dimension d is set by the WavLM.

---

[2] https://huggingface.co/microsoft/wavlm-large.

Physiologically Grounded Prosodic Features. Age-related changes in vocal production often manifest as measurable alterations in pitch variability, speech tempo, and pause behavior. These acoustic markers have been consistently linked to emotional flattening, cognitive slowing, and neuromuscular decline in elderly individuals [12]. To capture such physiologically relevant cues, we extract four segment-level prosodic features that serve as indicators of vocal expressiveness and temporal regulation.

(1) **Pitch Statistics.** Fundamental frequency (F_0) reflects vocal fold vibration and is a key correlate of arousal and vocal control [22]. In elderly individuals, reduced pitch variability is associated with affective flattening and physiological aging [33]. We extract frame-level $F_0^{i,t,s}$ using the YIN algorithm via `librosa.pyin`[3] with a frequency range of 50–500 Hz, and computes:

$$\mu_{F_0}^{t,s} = \frac{1}{N_{t,s}} \sum_{i=1}^{N_{t,s}} F_0^{i,t,s}, \quad \sigma_{F_0}^{t,s} = \sqrt{\frac{1}{N_{t,s}} \sum_{i=1}^{N_{t,s}} \left(F_0^{i,t,s} - \mu_{F_0}^{t,s} \right)^2}, \tag{5}$$

where $N_{t,s}$ is the number of voiced frames in segment s on day t.

(2) **Speech Rate.** Speech rate reflects cognitive fluency and emotional activation [21]. Slower or irregular pacing may indicate depressive symptoms or cognitive decline. We define:

$$\mathrm{SR}^{t,s} = \frac{N_{\mathrm{syll}}^{t,s}}{T_{\mathrm{speech}}^{t,s}}, \tag{6}$$

where $N_{\mathrm{syll}}^{t,s}$ is the estimated syllable count, and $T_{\mathrm{speech}}^{t,s}$ is the total non-silent duration. Syllables are detected as peaks in a low-pass-filtered energy envelope using `librosa.onset.onset_detect`[4], with a minimum inter-peak distance of 100 ms. $T_{\mathrm{speech}}^{t,s}$ is computed by summing all frames with energy exceeding 10% of the segment's maximum.

(3) **Silence Ratio.** Prolonged or frequent pauses reflect hesitation, cognitive load, or emotional withdrawal [26]. We compute the proportion of silent frames as:

$$R_{\mathrm{pause}}^{t,s} = \frac{1}{N_{t,s}} \sum_{i=1}^{N_{t,s}} \mathbb{I}(E_i^{t,s} < 0.1 \cdot \max_{1 \leq j \leq N_{t,s}} E_j^{t,s}), \tag{7}$$

where $E_i^{t,s} = \sum_{j=1}^{L} (r_{i,j}^{t,s})^2$ is the short-time energy of frame i, and $\mathbb{I}(\cdot)$ is the indicator function.

(4) **Jitter.** Jitter captures micro variations in pitch periods and is a clinical marker of vocal instability [34]. In elderly speech, elevated jitter is linked to reduced laryngeal control or neuromuscular impairment [4]. Let $T_i^{t,s}$ denote the

---

[3] https://librosa.org/doc/main/generated/librosa.pyin.html.
[4] https://librosa.org/doc/main/generated/librosa.onset.onset_detect.html.

pitch period at time i, extracted using `parselmouth`[5]. We compute:

$$J^{t,s} = \frac{1}{N_{t,s}-1} \sum_{i=1}^{N_{t,s}-1} \frac{|T_{i+1}^{t,s} - T_i^{t,s}|}{\frac{1}{N_{t,s}-1}\sum_{i=1}^{N_{t,s}-1} T_i^{t,s}}. \tag{8}$$

After computing these physiologically grounded prosodic features respectively, we concatenate them into a single segment-level prosodic vector $\mathbf{a}_{\mathrm{pro}}^{t,s} = [\mu_{F_0}^{t,s}, \sigma_{F_0}^{t,s}, \mathrm{SR}^{t,s}, R_{\mathrm{pause}}^{t,s}, J^{t,s}] \in \mathbb{R}^5$, which quantifies vocal expressiveness and timing irregularities associated with aging-related emotional decline.

Prosodic Gated Fusion. To combine the expressive capacity of pretrained embeddings with the physiologically grounded prosodic features, we introduce a learnable gating mechanism that dynamically balances the two sources at the segment level:

$$\mathbf{g}_{\mathrm{a}}^{t,s} = \sigma(W_g^{\mathrm{a}}[\mathbf{a}_{\mathrm{pre}}^{t,s} \| \mathbf{a}_{\mathrm{pro}}^{t,s}]+b_g^{\mathrm{a}}), \quad \mathbf{a}^{t,s} = \mathbf{g}_{\mathrm{a}}^{t,s}\odot\mathbf{a}_{\mathrm{pre}}^{t,s}+(1-\mathbf{g}_{\mathrm{a}}^{t,s})\odot(W_h^{\mathrm{a}}\mathbf{a}_{\mathrm{pro}}^{t,s}+b_h^{\mathrm{a}}). \tag{9}$$

Here, $\odot$ denotes element-wise multiplication, $\|$ indicates vector concatenation, and $(W_g^{\mathrm{a}}, b_g^{\mathrm{a}}, W_h^{\mathrm{a}}, b_h^{\mathrm{a}})$ are learnable parameters. Specifically, $W_g^{\mathrm{a}} \in \mathbb{R}^{d\times(d_1+d_2)}$, $W_h^{\mathrm{a}} \in \mathbb{R}^{d\times d_2}$, and $b_g^{\mathrm{a}}, b_h^{\mathrm{a}} \in \mathbb{R}^d$, where d_1 and d_2 are the dimensions of pretrained and prosodic features, respectively.

3.5 Linguistic Feature Extraction Module

Linguistic features in elderly speech, such as frequent self-reference, use of negative affective terms, and reduced cognitive word usage, provide valuable insight into emotional vulnerability and early cognitive decline [31,38]. These patterns are less affected by prosodic variation and complement acoustic features in spontaneous and low-resource settings [28]. To capture these linguistic signals effectively, CEDME combines two sources of information: (1) contextual embeddings from a pretrained BERT model, which encode semantic and syntactic structure, and (2) three psycholinguistic indicators that quantify the use of self-focused, negative, and cognitive language. This hybrid representation enables CEDME to capture both deep contextual semantics and psychologically salient lexical patterns. We describe each component in the following subsections.

Contextual Embedding. Given a segment on day t, denoted as the s-th utterance with token sequence $\{w_1^{t,s}, \ldots, w_{M_{t,s}}^{t,s}\}$, we obtain its contextual representation by encoding it with BERT-wwm[6] and extracting the final hidden state of the [CLS] token:

$$\mathbf{l}_{\mathrm{bert}}^{t,s} = \mathrm{BERT}([\mathrm{CLS}], w_1^{t,s}, \ldots, w_{M_{t,s}}^{t,s})_{[\mathrm{CLS}]} \in \mathbb{R}^{d_b}. \tag{10}$$

---

[5] https://parselmouth.readthedocs.io/.
[6] https://github.com/ymcui/Chinese-BERT-wwm/tree/master.

Psycholinguistic Indicators. To capture psychologically meaningful language patterns, we compute three segment-level psycholinguistic indicators, each reflecting an aspect of emotional or cognitive state in elderly speech. Specifically, FP (First-Person Pronouns) indicates inward attention and depressive tendencies [31]; NR (Negative Ratio) reflects emotional distress [36]; and CR (Cognitive Ratio) captures reasoning and reflective thinking [1]. Let $\tilde{\mathbf{w}}_t^s = \{\tilde{w}_1^{t,s}, \ldots, \tilde{w}_{\tilde{M}_{t,s}}^{t,s}\}$ denote the reconstructed word sequence obtained by merging contiguous subword tokens from $\mathbf{w}_t^s$. We define:

$$X^{t,s} = \frac{1}{\tilde{M}_{t,s}} \sum_{i=1}^{\tilde{M}_{t,s}} \mathbb{I}(\tilde{w}_i^{t,s} \in \mathcal{V}_Z), \quad (X, Z) \in \{(\text{FP}, \text{self}), (\text{NR}, \text{neg}), (\text{CR}, \text{cog})\}. \tag{11}$$

Here, $\tilde{M}_{t,s}$ is the number of reconstructed words in segment s on day t, and $\mathcal{V}_Z$ is the psychologically salient lexicon. Specifically, $\mathcal{V}_{\text{self}}$ and $\mathcal{V}_{\text{cog}}$ are derived from the Chinese LIWC dictionary[7], while $\mathcal{V}_{\text{neg}}$ is taken from cntext[8]. The final linguistic feature vector is $\mathbf{l}_{\text{lex}}^{t,s} = [\text{FP}^{t,s}, \text{NR}^{t,s}, \text{CR}^{t,s}] \in \mathbb{R}^3$.

Linguistic Gated Fusion. To integrate semantic embeddings with lexically grounded features, we apply a learnable gating mechanism at the segment level:

$$\mathbf{g}_\text{l}^{t,s} = \sigma\left(W_g^\text{l}[\mathbf{l}_{\text{bert}}^{t,s} \| \mathbf{l}_{\text{lex}}^{t,s}] + b_g^\text{l}\right), \quad \mathbf{l}^{t,s} = \mathbf{g}_\text{l}^{t,s} \odot \mathbf{l}_{\text{bert}}^{t,s} + (1 - \mathbf{g}_\text{l}^{t,s}) \odot \left(W_h^\text{l}\mathbf{l}_{\text{lex}}^{t,s} + b_h^\text{l}\right). \tag{12}$$

Here, $(W_g^\text{l}, b_g^\text{l}, W_h^\text{l}, b_h^\text{l})$ are learnable parameters. Specifically, $W_g^\text{l} \in \mathbb{R}^{d' \times (d_3+d_4)}$, $W_h^\text{l} \in \mathbb{R}^{d' \times d_4}$, and $b_g^\text{l}, b_h^\text{l} \in \mathbb{R}^{d'}$, where d_3 and d_4 denote the dimensions of the BERT embedding and psycholinguistic feature vector, respectively. Note that we set $d = d'$ to ensure alignment.

3.6 Hierarchical Multimodal Temporal Scoring Module

To model emotional trajectories over time, we employ a two-level hierarchical architecture that captures both within-day and across-day dynamics. At the segment level, we first combine the prosodic and linguistic representations via element-wise addition to form a unified multimodal vector: $\mathbf{x}^{t,s} = \mathbf{a}^{t,s} + \mathbf{l}^{t,s} \in \mathbb{R}^d$. Given the sequence of segments on day t, denoted as $\{\mathbf{x}^{t,1}, \ldots, \mathbf{x}^{t,S_t}\}$, we use a bidirectional LSTM [10] to encode short-term temporal context. The resulting forward and backward hidden states $\overrightarrow{\mathbf{h}}^{t,s}, \overleftarrow{\mathbf{h}}^{t,s} \in \mathbb{R}^h$ are concatenated to form the segment-level context vector $\mathbf{u}^{t,s} \in \mathbb{R}^{2h}$:

$$\mathbf{u}^{t,s} = \text{BiLSTM}(\mathbf{x}^{t,s}), \quad s = 1, \ldots, S_t. \tag{13}$$

To summarize variable-length segment sequences into a consistent representation for each day, we apply attention-based pooling to aggregate segment-level

---

[7] https://cliwceg.weebly.com/.
[8] https://github.com/hiDaDeng/cntext/.

context vectors $\mathbf{u}^{t,s}$ into a fixed-length daily embedding $\mathbf{e}^t$:

$$\alpha_s^t = \frac{\exp(\mathbf{f}^\top \tanh(U\mathbf{u}^{t,s}))}{\sum_{j=1}^{S_t} \exp(\mathbf{f}^\top \tanh(U\mathbf{u}^{t,j}))}, \quad \mathbf{e}^t = \sum_{s=1}^{S_t} \alpha_s^t \mathbf{u}^{t,s}. \tag{14}$$

Here, $U \in \mathbb{R}^{d_{attn} \times 2h}$ and $\mathbf{f} \in \mathbb{R}^{d_{attn}}$ are learnable parameters, and S_t is the number of segments on day t. To model emotional dynamics across multiple days, we process the daily representations $\mathbf{e}^t$ with a temporal Transformer encoder [17]. Each $\mathbf{e}^t$ is augmented with a learnable positional embedding $\mathbf{p}^t$, and the resulting sequence is processed by a stack of L standard Transformer layers:

$$\mathbf{h}_t^{(0)} = \mathbf{e}^t + \mathbf{p}^t, \quad \mathbf{h}_t^{(l)} = \text{TransformerLayer}^{(l)}(\mathbf{h}_t^{(l-1)}), \quad l = 1, \ldots, L. \tag{15}$$

To summarize emotionally informative cues across days while prioritizing recent sessions, we apply self-attention pooling with a time-aware decay bias:

$$\beta_t = \frac{\exp(\mathbf{q}^\top \tanh(W\mathbf{h}_t^{(L)}) - \gamma(T - t))}{\sum_{k=1}^{T} \exp(\mathbf{q}^\top \tanh(W\mathbf{h}_k^{(L)}) - \gamma(T - k))}, \quad \mathbf{v} = \sum_{t=1}^{T} \beta_t \mathbf{h}_t^{(L)}. \tag{16}$$

Here, $W \in \mathbb{R}^{d_q \times d}$ and $\mathbf{q} \in \mathbb{R}^{d_q}$ are trainable parameters, and $\gamma = 0.1$ is a fixed decay coefficient controlling the strength of temporal discounting. Unlike Eq.(14), which aggregates within-day segments, this formulation explicitly incorporates temporal distance to highlight recent emotional patterns while maintaining global context. The emotional well-being score $s \in \{1, \ldots, 10\}$ is then predicted by passing $\mathbf{v}$ through a feedforward network followed by softmax:

$$\mathbf{Z} = W_2 \, \text{ReLU}(W_1\mathbf{v} + b_1) + b_2, \quad P_i = \frac{\exp(Z_i)}{\sum_{j=1}^{10} \exp(Z_j)}, \quad i = 1, \ldots, 10. \tag{17}$$

The predicted probability distribution $\mathbf{P} = [P_1, \ldots, P_{10}]$ reflects the module's confidence across the 10 ordinal classes. The final well-being score is obtained by $s = \arg\max_{i \in \{1,\ldots,10\}} P_i$. While standard cross-entropy loss is commonly used for classification, it treats all misclassifications equally and does not consider the inherent ordering of score labels. For instance, predicting a score of 9 when the true label is 10 should be penalized less than predicting a score of 1. To incorporate this ordinal structure into the objective, we adopt a distance-aware weighted cross-entropy loss:

$$\mathcal{L} = -\frac{1}{N} \sum_{n=1}^{N} \sum_{i=1}^{10} \omega(|i - y_n|) \, y_{n,i} \log p_{n,i}, \quad \omega(d) = 1 + \zeta d^2, \tag{18}$$

where N is the total number of training samples, $y_n \in \{1, \ldots, 10\}$ is the true score label, $y_{n,i}$ its one-hot encoding, and ζ is empirically set to 0.5 to balance robustness and sensitivity to ordinal distance.

As shown in Table 1, we map the predicted emotional well-being score $s \in \{1, \ldots, 10\}$ to discrete care recommendations based on risk stratification

Table 1. Care recommendation mapping based on emotional well-being score.

Score s	Risk Level	Recommended Intervention
$s = 1$	Extreme	Urgent in-person visit and referral to mental health services
$s = 2$	Very High	Immediate phone call and same-week clinical evaluation
$s = 3$	High	Daily phone check-ins and close monitoring by care team
$s = 4$	Elevated	Phone or volunteer contact every other day
$s = 5$	Moderate	Twice-weekly contact and emotional status journaling
$s = 6$	Mild	Weekly contact and general wellness encouragement
$s = 7$	Low	Routine check-in every 10 d; optional automated reports
$s = 8$	Very Low	Bi-weekly greetings and passive emotional trend monitoring
$s = 9$	Stable	Monthly report with summary feedback
$s = 10$	Normal	No action required; maintain current contact frequency

practices in geriatric mental health [14]. The score range is grouped into three risk levels (low, moderate, and high), guided by thresholds in prior studies [7,27] and clinical tools such as the Geriatric Depression Scale (GDS-15) [27]. This mapping supports tiered interventions, from routine check-ins to urgent referrals.

(a) Face-to-face interactions   (b) Smart speakers

Fig. 2. Examples of conversational segment collection in the CESD dataset.

4   Experiments

• **Datasets.** We construct a Chinese elderly speech dataset (CESD) for this study. The data are collected from older adults aged over 70, including both male and female participants, living alone or in community-based care centers in Beijing and nearby regions. Participants are recruited through local health-care providers, with informed consent obtained. As shown in Fig. 2, each subject contributes 14 consecutive days of speech, consisting of brief conversational segments collected via phone calls, smart speakers, voice messages, and face-to-face interactions. Recordings are conducted in natural settings using portable devices. Each audio segment is annotated with speaker ID, gender, date index, and segment index within the day. The dataset is split into training and test sets. Final well-being scores are assigned by licensed psychologists based on multi-day speech observations, serving as the ground-truth labels for model supervision.

• **Baselines.** To the best of our knowledge, and despite exhaustive efforts, we find no existing non-intrusive, multimodal methods that enable fair and direct comparison for long-range emotional decline monitoring in elderly individuals. We therefore adopt general-purpose models and simplified heuristics as proxy baselines. (1) **GPT-4o** [23]: A multimodal model capable of processing speech and text. We prompt it with daily utterances and aggregate responses into well-being scores. (2) **Doubao** [6]: A multimodal model that processes raw audio and text. We prompt it with daily recordings and transcripts to infer emotional trends. (3) **Human Evaluators**: Ten non-author psychology-trained students listen to multi-day recordings and rate emotional well-being on a 1–10 scale. (4) **Rule-Based Estimator**: A rule-based method that adjusts the emotional well-being score based on trends in speech rate and pause frequency: the score decreases if both worsen for three days, and increases if both improve.

• **Setup.** CEDME is implemented in PyTorch and trained on a single `NVIDIA A100 GPU`. The BiLSTM encoder uses a hidden size of 256 per direction, and the daily-level Transformer has $L = 4$ layers with 8 heads and a hidden size of 512. We set $d_{attn} = 128$, $d = 256$, and apply dropout at 0.3. Training uses the Adam optimizer with a learning rate of 1e−4, weight decay of 5e−5, and batch size of 16 with each batch covering $T = 14$ days.

• **Evaluation Metrics.** We evaluate CEDME from four perspectives: (1) Prediction Performance: assessed using standard metrics for emotional well-being prediction. (2) Early Warning: evaluated via forward prediction on truncated inputs to test anticipatory capability. (3) Module Contribution: examined through ablation studies to assess the impact of key components. (4) Prosodic Explainability: analyzed using SHAP to interpret feature contributions.

4.1   Performance Evaluation

As shown in Table 2, we evaluate the performance of CEDME and four baselines on the CESD dataset using three standard metrics for multi-class emotional prediction: unweighted average recall (UAR), macro-averaged F1 score (Macro-F1), and linearly weighted Cohen's Kappa (Kappa) [5]. CEDME consistently outperforms all baselines, achieving a UAR of 64.6%, which is 6.5% higher than GPT-4o, along with 61.4% Macro-F1 and 0.563 Kappa. These results confirm CEDME's strength in detecting emotional changes from sparse and non-intrusive daily speech. Compared to Doubao and GPT-4o, which rely on prompt-based snapshot inference, CEDME models both intra-day and inter-day dynamics through hierarchical temporal structure, and benefits from robust multimodal fusion of prosodic and linguistic cues.

Table 2. Performance evaluation on dataset CESD.

Model	UAR (%)	Macro-F1 (%)	Kappa
Rule-Based Estimator	32.3	29.5	0.262
Human Evaluators	55.6	51.2	0.459
Doubao	53.7	49.8	0.441
GPT-4o	58.1	55.3	0.497
CEDME (Ours)	**64.6**	**61.4**	**0.563**

4.2 Early Warning Capability

To evaluate CEDME's ability to anticipate emotional decline, we conduct a forward prediction experiment by truncating the input sequence at $T-k$ days and predicting the emotional well-being score on day T, with $k = 5, 3, 1$ representing different early-warning horizons. We consider two tasks: regression, using mean absolute error (MAE) to evaluate predicted scores s_T, and binary classification, identifying high-risk cases ($s_T \leq 3$) using area under the precision-recall curve (AUC-PR). As shown in Fig. 3, CEDME achieves progressively lower MAE and higher AUC-PR as the input window approaches day T, indicating improved prediction accuracy and sensitivity. These results suggest that CEDME captures meaningful early indicators of emotional decline from past speech.

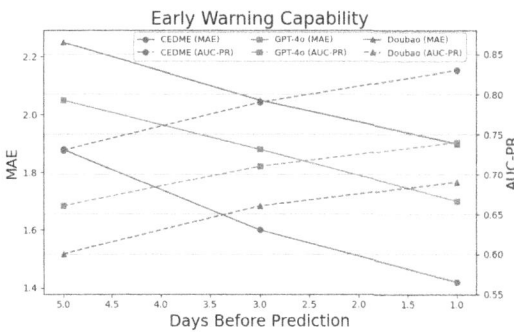

Fig. 3. The Early Warning Capability of CEDME and baselines.

4.3 Ablation Study

As shown in Table 3, we conduct an ablation study to assess the contribution of each core component within CEDME. The tested variants are: (1) **CEDME w/o Cross-Day Transformer**, which removes the cross-day Transformer encoder and relies solely on the segment-level BiLSTM; (2) **CEDME (ProsodyOnly)** and **CEDME (TextOnly)**, which retain only the prosodic branch or the linguistic branch, respectively; (3) **CEDME w/o GatedFusion**, which replaces the gated multimodal fusion with naive concatenation followed by a fully connected layer; and (4) **CEDME w/o DecayBias**, which removes the time decay bias γ from the cross-day attention mechanism. Results indicate that each component is critical to overall performance. The cross-day Transformer contributes most significantly, underscoring the value of long-range temporal modeling. Gated fusion and time-aware scoring offer measurable gains, and both prosodic and textual features are essential for robust prediction.

Table 3. Ablation study on CEDME. Each variant shows performance drop.

Model Variant	ΔUAR (%)	ΔMacro-F1 (%)	ΔKappa
CEDME w/o Cross-Day Transformer	−6.8	−6.0	−0.081
CEDME (ProsodyOnly)	−6.1	−5.2	−0.073
CEDME (TextOnly)	−4.8	−4.1	−0.061
CEDME w/o GatedFusion	−3.1	−2.6	−0.033
CEDME w/o Time Decay Bias	−2.2	−1.7	−0.021

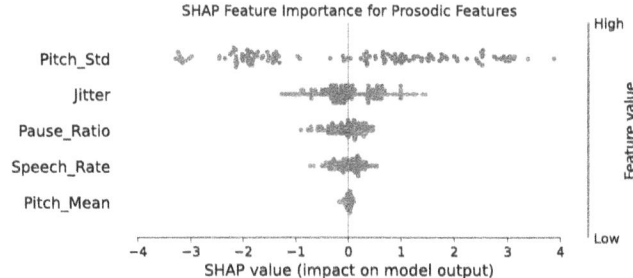

Fig. 4. The SHAP importance of five prosodic features used in CEDME.

4.4   Prosodic Feature Explainability via SHAP

To enhance the interpretability and clinical trustworthiness of CEDME, we perform post-hoc attribution analysis using SHAP [18]. Specifically, we apply SHAP to the prosodic gating module, focusing on the five physiologically grounded features: $\mu_{F_0}^{t,s}$ (Pitch_Mean), $\sigma_{F_0}^{t,s}$ (Pitch_Std), $SR^{t,s}$ (Speech_Rate), $R_{\text{pause}}^{t,s}$ (Pause_Ratio), and $J^{t,s}$ (Jitter). We visualize global importance using SHAP summary plots and highlight local decision factors through force plots for individual users. As shown in Fig. 4, features such as lower pitch and higher pause ratio consistently contribute to lower predicted scores. This aligns with speech pathology literature, where reduced pitch variability and prolonged silences are markers of affective flattening in older adults. These results confirm that CEDME not only performs well but also makes predictions aligned with known vocal indicators of emotional decline.

5   Conclusion

In this work, we present CEDME, a multimodal and temporally structured framework for continuous emotional decline monitoring in the Chinese elderly. By integrating prosodic cues with linguistic features, CEDME captures fine-grained affective dynamics across both intra-day and across-day scales. Extensive experiments demonstrate that CEDME outperforms strong baselines, including human evaluators and state-of-the-art multimodal models, in terms of accuracy, early-warning capability, and explainability.

References

1. Badal, V., Nebeker, C., Yamada, Y., Rentscher, K., Kim, H.C., Lee, E.: Do words matter? detecting social isolation and loneliness in older adults using natural language processing. Front. Psychiatry 728732 (2021)
2. Boateng, G., Kowatsch, T.: Speech emotion recognition among elderly individuals using multimodal fusion and transfer learning. In: Proceedings of ICMI, pp. 12–16 (2021)
3. Canali, S., Ferretti, A., Schiaffonati, V., Blasimme, A.: Wearable technologies for healthy ageing: Prospects, challenges, and ethical considerations. J. Frailty Aging (2024)
4. Cavallaro, G., Fiorella, M., Barbara, F., Quaranta, N., Di Nicola, V.: Exploring age-related changes in acoustic voice analysis parameters: insights from a study on older people. J. Gerontol. Geriatr. 1–9 (2024)
5. Cohen, J.: Weighted kappa - nominal scale agreement with provision for scaled disagreement or partial credit. Psychol. Bull. 213–220 (1968)
6. Doubao Team: Doubao 1.5 pro technical report (2024). https://seed.bytedance.com/en/special/doubao_1_5_pro
7. Evans, L., et al.: Risk stratification models for predicting preventable hospitalization in commercially insured late middle-aged adults with depression. BMC Health Serv. Res. (2023)
8. Feng, G., Wang, H., Wang, M., Zheng, X., Zhang, R.: A research on emotion recognition of the elderly based on transformer and physiological signals. Electronics (2024)
9. Finze, N., Jechle, D., Faußer, S., Gewald, H.: How are we doing today? using natural speech analysis to assess older adults' subjective well-being. Bus. Inf. Syst. Eng. 321–334 (2024)
10. Graves, A., Schmidhuber, J.: 2005 special issue: Framewise phoneme classification with bidirectional LSTM and other neural network architectures. Neural Netw. 602–610 (2005)
11. Guo, R., Guo, H., Wang, L., Chen, M., Yang, D., Li, B.: Development and application of emotion recognition technology – a systematic literature review. BMC Psychol. (2024)

12. Harlev, D., Singer, S., Goldshalger, M., Wolpe, N., Bergmann, E.: Acoustic speech features are associated with late-life depression and apathy symptoms: Preliminary findings. Alzheimers Dement. Diagn. Assess. Dis. Monit. e70055 (2025)

13. Hou, M., Zhang, X., Chen, G., Huang, L., Sun, Y.: Emotion recognition based on a EEG–FNIRs hybrid brain network in the source space. Brain Sci. 1166 (2024)

14. Kovacs, N., Biro, E., Piko, P., Ungvari, Z., Adany, R.: Age-related shifts in mental health determinants from a deprived area in the european union: informing the national healthy aging program of hungary. GeroScience (2024)

15. Lei, J., Siriaraya, P., Choi, D., Kuwahara, N.: Emotion Recognition Using Electroencephalography Signals of Older People For Reminiscence Therapy. Front, Physiol (2022)

16. Li, X.: Evaluation and analysis of elderly mental health based on artificial intelligence. Occup. Ther. Int. 1–11 (2023)

17. Lim, B., Arik, S.O., Loeff, N., Pfister, T.: Temporal fusion transformers for interpretable multi-horizon time series forecasting (2020)

18. Lundberg, S.M., Lee, S.I.: A unified approach to interpreting model predictions. In: Proceedings of NeurIPS, pp. 4768–4777 (2017)

19. Maestro, E.G., de Almeida, T.R., Schaffernicht, E.J., Mozos, Ó.M.: Wearable-Based Intelligent Emotion Monitoring in Older Adults During Daily Life Activities. Appl. Sci. (2023)

20. Mu, X., et al.: Detecting cognitive impairment and psychological well-being among older adults using facial, acoustic, linguistic, and cardiovascular patterns derived from remote conversations (2025)

21. Neff, P., Demiray, B., Martin, M., Röcke, C.: Cognitive Abilities Predict Naturalistic Speech Length in Older Adults. Sci, Rep (2024)

22. Nussbaum, C., Schirmer, A., Schweinberger, S.: Contributions of Fundamental Frequency and Timbre to Vocal Emotion Perception and their Electrophysiological Correlates. Soc. Cogn. Affect, Neurosci (2022)

23. OpenAI: Gpt-4o system card (2024)

24. Padma, S., Veni, S., Murthy, O.: Elder emotion classification through multimodal fusion of intermediate layers and cross-modal transfer learning. Signal Image Video Process. 1281–1288 (2022)

25. Palmero, C., et al.: Exploring emotion expression recognition in older adults interacting with a virtual coach (2023)

26. Pastoriza-Domínguez, P., et al.: Speech pause distribution as an early marker for Alzheimer's disease. Speech Commun. 107–117 (2022)

27. Physiopedia: Geriatric depression scale — physiopedia, (2023)

28. Richard, A.B., Lelandais, M., Reilly, K.T., Jacquin-Courtois, S.: Linguistic markers of subtle cognitive impairment in connected speech: A systematic review. J. Speech Lang. Hear. Res. 4714–4733 (2024)

29. Ross, E.D.: Affective prosody and its impact on the neurology of language, depression, memory and emotions. Brain Sci. 1572 (2023)

30. Souganciouglu, G., Verkholyak, O., Kaya, H., Fedotov, D., Cadee, T., Salah, A.A., Karpov, A.: Is everything fine, grandma? Acoustic and linguistic modeling for robust elderly speech emotion recognition. In: Interspeech (2020)

31. Stade, E., Ungar, L., Eichstaedt, J., Sherman, G., Ruscio, A.: Depression and Anxiety Have Distinct and Overlapping Language Patterns: Results From a Clinical Interview. J. Psychopathol. Clin, Sci (2023)

32. Stuldreher, I.V., Thammasan, N., van Erp, J., Brouwer, A.M.: Physiological Synchrony in EEG, Electrodermal Activity and Heart Rate Detects Attentionally Relevant Events in Time. Front, Neurosci (2020)

33. Taylor, S., Dromey, C., Nissen, S., Tanner, K., Eggett, D., Corbin-Lewis, K.: Age-related changes in speech and voice: Spectral and cepstral measures. J. Speech Lang. Hear. Res. 647–660 (2020)

34. Teixeira, J., Oliveira, C., Lopes, C.: Vocal acoustic analysis – jitter, shimmer and hnr parameters. Procedia Technol. 1112–1122 (2013)

35. Wolf, A., Tripanpitak, K., Umeda, S., Otake-Matsuura, M.: Eye-tracking paradigms for the assessment of mild cognitive impairment: a systematic review. Front. Psychol. 1197567 (2023)

36. Yin, J., John, A., Cadar, D.: Bidirectional associations of depressive symptoms and cognitive function over time. JAMA Netw. Open e2416305–e2416305 (2024)

37. Zhao, L.: China's aging population: A review of living arrangement, intergenerational support, and wellbeing. Health Care Sci. 317–327 (2023)

38. Časnochová Zozuk, N.: Lexical diversity and language impairment. J. Linguist. 301–309 (2023)

Rehabilitation Exercise Quality Assessment and Feedback Generation Using Large Language Models with Prompt Engineering

Jessica Tang[1,2], Ali Abedi[1(✉)], Tracey J.F. Colella[1],
and Shehroz S. Khan[1,3]

[1] KITE Research Institute, Toronto Rehabilitation Institute,
University Health Network, Toronto, Canada
`jessicao.tang@mail.utoronto.ca,`
`{ali.abedi,tracey.colella,shehroz.khan}@uhn.ca`
[2] Faculty of Applied Science and Engineering, University of Toronto,
Toronto, Canada
[3] College of Engineering and Technology, American University of the Middle East,
Egaila, Kuwait

Abstract. Exercise-based rehabilitation improves quality of life and reduces morbidity, mortality, and rehospitalization, though transportation constraints and staff shortages lead to high dropout rates from rehabilitation programs. Virtual platforms enable patients to complete prescribed exercises at home, while AI algorithms analyze performance, deliver feedback, and update clinicians. Although many studies have developed machine learning and deep learning models for exercise quality assessment, few have explored the use of large language models (LLMs) for feedback and are limited by the lack of rehabilitation datasets containing textual feedback. In this paper, we propose a new method in which exercise-specific features are extracted from the skeletal joints of patients performing rehabilitation exercises and fed into pre-trained LLMs. Using a range of prompting techniques, such as zero-shot, few-shot, chain-of-thought, and role-play prompting, LLMs are leveraged to evaluate exercise quality and provide feedback in natural language to help patients improve their movements. The method was evaluated through extensive experiments on two publicly available rehabilitation exercise assessment datasets (UI-PRMD and REHAB24-6) and showed promising results in exercise assessment, reasoning, and feedback generation. This approach can be integrated into virtual rehabilitation platforms to help patients perform exercises correctly, support recovery, and improve health outcomes.

Keywords: Rehabilitation Exercise · Action Quality Assessment · Large Language Models · Feedback Generation · Prompt Engineering

1 Introduction

Patients recovering from a cardiac event, stroke, or other traumatic injuries are often referred to rehabilitation programs to support faster recovery. These programs aim to enhance patients' quality of life by promoting independent living

© The Author(s), under exclusive license to Springer Nature Singapore Pte Ltd. 2025
S. S. Khan et al. (Eds.): IJCAI 2025, CCIS 2620, pp. 60–75, 2025.
https://doi.org/10.1007/978-981-95-0568-5_5

and reducing the likelihood of hospital readmissions, morbidity, and mortality [1]. Typically, rehabilitation programs focus on exercises designed to restore mobility, rebuild muscle mass, and improve overall strength [2]. Traditionally delivered in clinical settings or institutional environments, these programs frequently face challenges such as long wait times, limited staffing, and logistical barriers to participation, including transportation difficulties and scheduling constraints [3]. Virtual and home-based rehabilitation programs [4] offer a practical alternative, addressing these challenges while providing benefits comparable to in-person care [5,6]. By analyzing data collected during virtual sessions, Artificial Intelligence (AI) can be used to assess exercise quality, monitor patient progress, and predict program dropout risks [7]. These AI-driven methods typically leverage various sensors, such as wearable devices and cameras, to track patient movements. AI algorithms process this information in real time [4,8], offering valuable insights into exercise performance and enabling healthcare professionals to effectively monitor and personalize patient care interventions [7,8].

AI-driven methods for assessing the quality of rehabilitation exercises primarily rely on three types of data: acceleration data from inertial wearable sensors, video data from RGB or depth cameras, and body joint data [8–10]. Body joint data, in particular, is either captured using sensors such as Kinect or extracted from RGB videos using computer vision techniques [8,11–13]. Prior research in general human activity analysis has emphasized the importance of body joint analysis in enhancing performance recognition [14]. In the context of rehabilitation, body joint analysis closely aligns with clinical practices used to evaluate exercise technique and quality [15]. Moreover, compared to video data, body joint data has lower dimensionality and is less affected by variations in lighting and background, making it a more robust and reliable modality for analysis. This paper focuses on evaluating rehabilitation exercises through the analysis of body joint sequences.

Real-time feedback on rehabilitation exercise quality and technique [9,10] plays a crucial role in enhancing the effectiveness of exercise-based rehabilitation. It not only helps correct movement execution and ensure proper technique but also motivates patients and strengthens their psychological resilience [7,16,17]. While many studies have explored body jointâĂŞbased assessment of exercise quality [8] and feedback generation [9,10], the potential of Large Language Models (LLMs) to deliver rich, personalized, and context-aware feedback remains largely underexplored [18].

This paper investigates the applicability of Large Language Models (LLMs) for assessing rehabilitation exercise quality and generating feedback, making the following key contributions: (1) it proposes a novel framework that enables pre-trained LLMs to evaluate rehabilitation exercise quality by integrating exercise-specific features with prompt engineering techniques; and (2) it conducts extensive experiments on two publicly available rehabilitation exercise datasets, the University of Idaho-Physical Rehabilitation Movements Dataset (UI-PRMD) [19] and the REHAB24-6 dataset [20], demonstrating the feasibility of using pre-trained LLMs for both exercise quality assessment and natural language feedback generation.

2 Related Work

This section briefly reviews existing approaches for rehabilitation exercise quality assessment [8] and feedback generation on exercise quality [9,10].

2.1 Rehabilitation Exercise Quality Assessment

Liao et al. [21] assessed rehabilitation exercise quality using principal component analysis for dimensionality reduction and Long Short-Term Memory (LSTM) autoencoders. Their model employed temporal pyramid sub-networks with 1D convolutions on joint sequences at different time resolutions, followed by LSTM layers for quality evaluation.

Abedi et al. [13] proposed a rehabilitation exercise quality assessment method using MediaPipe for joint extraction, followed by LSTM models trained on exercise-specific features [22]. To enhance generalizability, they applied cross-modal video-to-body-joint augmentation. Karagoz et al. [23] further improved this approach with supervised contrastive learning to address imbalanced exercise sample distributions.

Deb et al. [24] applied Spatial-Temporal Graph Convolutional Networks (ST-GCNs) for rehabilitation exercise quality assessment, replacing global average pooling with an LSTM layer to enhance performance. Zheng et al. [25] improved the robustness of ST-GCNs by introducing a rotation-invariant descriptor. Réby et al. [26] combined ST-GCNs with transformers by incorporating spatial and temporal self-attention modules, though their approach did not outperform the baseline ST-GCN model [14]. Karlov et al. [27] further advanced ST-GCNs by integrating contrastive learning with hard and soft negatives to develop an exercise-agnostic model for more effective assessment. Most recently, Bruce et al. [28] proposed an Ensemble-based Graph Convolutional Network (EGCN++), which integrates position and orientation features through a novel fusion strategy, achieving improved performance in rehabilitation exercise quality assessment.

Apart from deep learning-based methods, threshold-based approaches have also been used for rehabilitation exercise quality assessment; however, they generally underperform compared to deep learning models due to their limited ability to handle variability in human movement and sensor noise [9]. While these methods can evaluate exercise quality through classification or regression (i.e., by assigning a class label or quality score), they do not generate natural language feedback to guide patients in correcting their exercise performance.

2.2 Rehabilitation Exercise Feedback Generation

Brennan et al. [10] and Ettefagh and Roshan Fekr [9] reviewed AI-driven rehabilitation exercise feedback and found most studies used inertial sensor data, with some leveraging Kinect-captured body joint data. Feedback was primarily delivered via animated avatars through audio, visual, haptic, or multimodal channels. Feedback was categorized into knowledge of results (e.g., repetition count) and knowledge of performance, which can be descriptive (error explanation) or prescriptive (correction guidance) [29]. However, neither review identified the use of LLMs for generating rehabilitation exercise feedback.

Wang et al. [18] introduced the first LLM-enabled platform for rehabilitation exercise feedback generation. They collected data from 104 subjects using inertial sensors, annotated with physiotherapist-guided corrections. Features extracted from the data, combined with an action token and a prompt, were used to fine-tune LLMs for feedback generation. Among the reviewed works, Wang et al. [18] is most closely related to the present study, as it represents an initial attempt to apply LLMs for rehabilitation exercise feedback. However, their approach required resource-intensive fine-tuning due to the lack of advanced prompting strategies, limiting its scalability and practical deployment. In contrast, the method proposed in this paper addresses this gap by demonstrating that strategically designed prompting can effectively enable LLMs to generate rehabilitation feedback without the need for costly model fine-tuning, thereby offering a more efficient and accessible solution.

3   Method

Figure 1 illustrates the proposed method. The input consists of a body joint sequence representing a subject performing one repetition of a rehabilitation exercise. The output includes an assessment of exercise quality and textual feedback on the performed movements. If the movements are incorrect, the feedback provides guidance on how to correct them to ensure proper execution of the exercise. Either the raw body joint sequence or a set of exercise-specific features extracted from the joints, along with a prompt and exercise type, is fed into a pre-trained LLM to generate both the quality assessment and the corresponding feedback.

The dimensionality of body joint sequences varies depending on the acquisition device. For example, in the UI-PRMD dataset [19], body joint data were extracted using a Kinect depth camera, capturing 22 joints with three spatial coordinates (x, y, z). Similarly, the REHAB24-6 dataset [20], which employed multiple wearable inertial measurement units, contains data for 26 joints, each with three channels (x, y, z). Each data sample represents a single repetition of an exercise and has a dimensionality of $num\_frames \times num\_joints \times num\_channels$.

3.1   Feature Extraction

The rehabilitation clinicians involved in designing the rehabilitation exercise program provided guidelines on how the exercises should be performed correctly and established criteria for identifying incorrect movements [15,19,20,30,31]. These

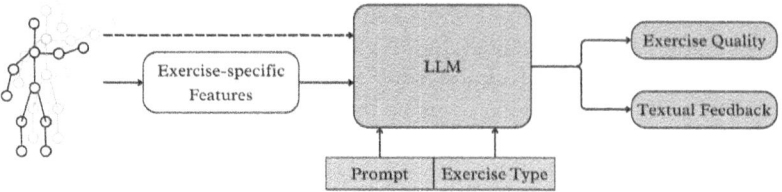

Fig. 1. Either body joint data or exercise-specific features extracted from body joints, combined with engineered prompts and exercise type, are fed into a pre-trained LLM. The LLM then generates exercise quality assessments and provides textual feedback.

guidelines and criteria were incorporated into the proposed method to extract exercise-specific features from body joint sequences. For instance, in the side lunge rehabilitation exercise in UI-PRMD, the correct movement is defined as "the subject takes a step to the side and lowers the body toward the floor" [19]. The non-optimal movements classified as incorrect include: "moderate to significant knee valgus collapse, pelvis dropping or rising more than 5°, trunk angle less than 30°, thigh angle exceeding 45°, and the center of the knee positioned anterior to the toes" [19]. Based on these criteria, the proposed method extracted three features from body joints at each frame: knee valgus angle, thigh angle, and pelvic stability. This resulted in a $num\_frames \times num\_features$ array, where $num\_features$ is 3. Information about the patients' dominant side was incorporated into the feature extractor for movements that are performed one side at a time. Tables 1 (a) and (b) outline the rehabilitation exercises included in UI-PRMD and REHAB24-6, respectively, along with three to five exercise-specific features extracted for each exercise as part of the proposed method. The code[1] released with this paper performs feature extraction from these datasets.

3.2 Prompt Engineering

A variety of prompting techniques [32] guide the LLM in analyzing feature sequences derived from rehabilitation exercise body joint data, assessing exercise quality, and generating textual feedback. The initial prompt, aimed at *classifying* exercises as correct and incorrect exercises, served as the basis for more advanced prompts that were subsequently developed (Table 2).

The classification prompt was initially evaluated across different shot settings, ranging from *zero-shot* to *few-shot* prompting [33]. Also known as *k-shot* prompting, this approach refers to the number of labeled examples (k) provided per class in classification tasks. In *zero-shot* prompting ($k = 0$), the LLM is presented with only a task description or question, relying solely on its pre-trained knowledge to generate a response. In contrast, *few-shot* prompting ($k \in \{1, 2, 3, 4\}$) improves performance by incorporating a small set of labeled examples into the input prompt, allowing the model to infer the task's structure

---

[1] https://github.com/jessicaxtang/exercisellm.

Table 1. Rehabilitation exercises from (a) UI-PRMD (m01–m10) and (b) REHAB24-6 (ex1–ex6) datasets with exercise-specific feature descriptions ("A." denotes Angle).

(a) UI-PRMD

#	Exercise	Extracted Features
m01	Deep squat	Knee Flexion A., Hip Flexion A., Trunk Inclination A.
m02	Hurdle step	Trunk Inclination A., Hip Flexion A., Leg Height
m03	Inline lunge	Front Knee A., Back Knee A., Trunk Inclination A., Foot Distance
m04	Side lunge	Knee Valgus A., Thigh A., Pelvic Stability
m05	Sit to stand	Trunk Inclination A., Hip Flexion A., Pelvic Stability
m06	Active straight leg raise	Hip Flexion A., Leg Elevation A., Pelvic Stability
m07	Shoulder abduction	Arm Elevation A., Elbow Flexion A., Torso Inclination A.
m08	Shoulder extension	Shoulder Extension A., Head Neutral Position, Trunk Inclination A.
m09	Shoulder internal-external rotation	Arm Internal Rotation A., Arm External Rotation A., Elbow Flexion A.
m10	Shoulder scaption	Arm Elevation A., Trunk Inclination A., Arm Plane Deviation

(b) REHAB24-6

#	Movement	Extracted Features
ex1	Arm Abduction	Arm Elevation A., Trunk Inclination A., Elbow A., Plane Deviation
ex2	Arm VW	V-Shape A. (shoulder), W-Shape A. (elbow), Trunk to Vertical A.
ex3	(Inclined) Push-ups	Elbow Flexion A., Trunk Inclination A., Hand Symmetry, Pelvic Stability
ex4	Leg Abduction	Leg Elevation A., Trunk A., Pelvic Tilt A., Knee A., Leg Plane Deviation
ex5	Leg Lunge	Front Knee A., Back Knee A., Trunk A., Foot Distance
ex6	Squats	Knee Flexion A., Hip Flexion A., Trunk A., Foot Symmetry

and apply it effectively [32,34]. The optimal k-shot setting, which achieved the highest exercise quality classification accuracy, was used in the following more advanced prompting techniques.

Reasoning-elicitation techniques are known to enhance deep-learning model performance [35] and can be categorized into two approaches: *white-box* and *black-box*. White-box methods access model weights or activations but are impractical for powerful yet closed-source LLMs. In contrast, *black-box* approaches involve prompting an LLM to articulate its reasoning [36] through techniques such as Chain-of-Thought (CoT) [37], certainty [35], and probability prompting [38], offering insights into the model's rationale behind classification. *CoT prompting* guides the LLM to break down complex tasks into intermediate reasoning steps, enabling a more systematic and accurate problem-solving process while also offering insights into the model's thought process. *Certainty elicitation* generates a numerical certainty score ranging from 0 to 1, alongside assessing exercise quality, indicating the model's confidence in the accuracy of its evaluation. *Probability elicitation* outputs the probability that the exercise is correct, quantifying exercise quality and allowing flexibility in setting thresholds to balance recall and precision trade-offs [38].

Table 2. Prompting techniques for rehabilitation exercise quality assessment. Example of 2-shot prompting for the squat exercise, where four labeled examples (two correct, two incorrect) are provided to the LLM to evaluate a 5th test sample. Data sequences follow each prompt.

Technique	Prompt
Classification	Identify the label for the **5th** data sample below containing sequences of features extracted from Kinect data of the squat exercise. Ensure the output adheres to the output format: "Label". The label is either 'correct' or 'incorrect'. <Data 1, Label 1: correct> ... <Data 4, Label 4: incorrect> <Data 5>
Chain-of-Thought	Identify the label for the **5th** data sample below containing sequences of features extracted from Kinect data of the squat exercise. Ensure the output adheres to the output format: "Label, Reasoning". Explain your reasoning step by step. <Data 1, Label 1: correct> ... <Data 4, Label 4: incorrect> <Data 5>
Probability	Identify the label for the **5th** data sample below containing sequences of features extracted from Kinect data of the squat exercise. Ensure the output adheres to the output format: "Probability". Provide a probability score, where a higher score means a higher probability towards 'correct' and a lower score for 'incorrect'. <Data 1, Label 1: correct> ... <Data 4, Label 4: incorrect> <Data 5>
Certainty	Identify the label for the **5th** data sample below containing sequences of features extracted from Kinect data of the squat exercise. Ensure the output adheres to the output format: "Label, Certainty". Give a score between 0 and 1 for how certain you are in your classification. <Data 1, Label 1: correct> ... <Data 4, Label 4: incorrect> <Data 5>

Role-play Prompting [39] was deployed to generate textual feedback by asking the LLM to complete the task while embodying a persona, such as a physiotherapist, to deliver succinct advice to improve exercise quality. Figure 3 shows an example of role-play prompts and corresponding textual feedback.

3.3 Evaluation Metrics

The performance of the proposed method for rehabilitation exercise quality assessment was quantitatively evaluated using the binary ground-truth labels provided in the two aforementioned datasets. The evaluation metrics used were accuracy, precision, recall, and F1 score for all prompting techniques, and also Area Under the Receiver Operating Characteristic Curve (AUC-ROC), Area Under the Precision-Recall Curve (AUC-PR) if probability was elicited. As no publicly available rehabilitation exercise dataset includes ground-truth textual feedback [8–10], the performance of the proposed method for feedback generation was assessed qualitatively.

4   Experiments

This section presents a quantitative evaluation of the proposed method for rehabilitation exercise quality assessment and a qualitative evaluation of textual feedback generation. The results are presented on two publicly available datasets REHAB24-6 and UI-PRMD. GPT-4o [40] was used as the pre-trained LLM in all experiments.

4.1   Datasets

UI-PRMD comprises 10 rehabilitation exercise types performed by 10 healthy subjects. Each subject completed 10 repetitions of each exercise, both correctly and incorrectly, on their dominant side. Body-joint data were captured using a Kinect sensor at 30 frames per second, with dataset samples corresponding to individual exercise repetitions. The 10 exercise types in UI-PRMD are outlined in Table 1 (a).

REHAB24-6 features data from 10 subjects performing 6 rehabilitation exercises, correctly and incorrectly. Body-joint data were collected using inertial wearable sensors. The dataset includes annotations on exercise correctness (correct vs. incorrect) and the start and end of exercise repetitions, enabling the creation of individual exercise repetition data samples. The 6 exercise types in REHAB24-6 are outlined in Table 1 (b).

4.2   Experimental Results

Few-Shot Prompting. Figure 2 illustrates the accuracy of LLM-based exercise quality classification as the number of labeled examples increases, while Table 3 presents the corresponding precision, recall, and F1 scores. It is evident that *zero-shot prompting* yields the lowest accuracy, and incorporating more examples into the prompt generally enhances performance. However, beyond three-shot prompting, performance becomes inconsistent and may even decline. This observation aligns with previous findings [33] that indicate that larger values of k does not always lead to improved outcomes in LLMs. Consequently, few-shot prompting proves to be well-suited to the task of exercise assessment, particularly when datasets are limited in size. Unlike traditional machine learning approaches that require large training sets, few-shot prompting achieves comparable results with only a small number of examples.

Furthermore, Fig. 2 highlights the impact of input data type on classification accuracy. The results indicate that using exercise-specific features extracted from body joint data yields superior performance compared to using only raw body joint data. This improvement is due to the fact that extracted features are more interpretable and encapsulate domain-specific knowledge, as outlined in the feature extraction process detailed in Sect. 3.1. Additionally, extracted features have a significantly lower dimensionality than raw body joint data, effectively abstracting less relevant joints for a given movement.

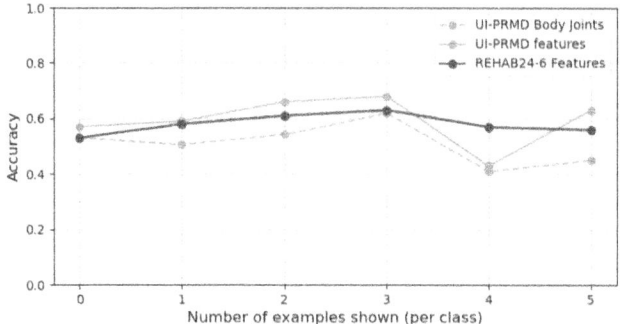

Fig. 2. Rehabilitation exercise quality classification accuracy varies with the number of labeled examples included in the prompts, evaluated on body joint data from UI-PRMD (orange dashed line), feature sequences extracted from UI-PRMD (orange solid line), and feature sequences extracted from REHAB24-6 (blue solid line). (Color figure online)

Table 3. Rehabilitation exercise quality classification accuracy, precision, recall, and F1 score varies with the number of labeled examples included in the prompts provided to the LLM, evaluated on feature sequences extracted from (a) UI-PRMD and (b) REHAB24-6. Bolded values denote the best results.

(a) UI-PRMD

	Accuracy	Precision	Recall	F1
0-shot	0.57	0.55	0.75	0.64
1-shot	0.59	0.57	0.72	0.63
2-shot	0.66	0.62	**0.84**	0.71
3-shot	**0.68**	**0.74**	0.79	**0.76**
4-shot	0.42	0.43	0.50	0.46
5-shot	0.63	0.64	0.60	0.62

(b) REHAB24-6

	Accuracy	Precision	Recall	F1
0-shot	0.53	0.54	0.73	0.62
1-shot	0.58	0.58	0.77	0.66
2-shot	0.61	0.59	0.81	0.68
3-shot	**0.63**	0.60	**0.85**	**0.70**
4-shot	0.57	**0.68**	0.72	0.65
5-shot	0.56	0.62	0.61	0.60

Reasoning Elicitation. Building on the best-performing setting from the few-shot prompting experiments, three-shot prompting with feature sequences was selected for reasoning elicitation and subsequent experiments. CoT, certainty, probability, and a combination of CoT-with-certainty were evaluated on the two datasets. As shown in Table 4, reasoning-elicitation methods generally outperformed baseline prompting. Among these methods, CoT and certainty prompting emerged as the most effective strategies. Notably, while accuracy scores remained similar for both approaches, slight differences were observed in precision and recall.

Table 4. Performance of the LLM with different prompting techniques on (a) UI-PRMD and (b) REHAB24-6. The results of LSTM and ST-GCN are presented for (b) REHAB24-6. Among the prompting techniques, only probability elicitation can generate class probability estimates, allowing for the computation of AUC-ROC and AUC-PR.

(a) UI-PRMD

Setting	Accuracy	Precision	Recall	F1	AUC-ROC	AUC-PR
3-shot	0.68	0.74	0.79	0.76	-	-
Chain-of-Thought	0.72	0.75	0.67	0.71	-	-
Certainty	0.76	0.72	0.87	0.79	-	-
Probability	0.68	0.65	0.79	0.71	0.70	0.68
Chain-of-Thought + Certainty	0.64	0.59	**0.90**	0.72	-	-
LSTM [13]	0.87	0.97	0.88	0.92	0.97	0.96
ST-GCN [25]	**0.94**	**0.98**	**0.90**	**0.96**	**0.98**	**0.98**

(b) REHAB24-6

Setting	Accuracy	Precision	Recall	F1	AUC-ROC	AUC-PR
3-shot	0.63	0.60	0.85	0.70	-	-
Chain-of-Thought	**0.70**	**0.71**	0.67	0.69	-	-
Certainty	**0.70**	0.67	**0.80**	**0.73**	-	-
Probability	0.67	0.63	**0.80**	0.71	0.72	0.68
Chain-of-Thought + Certainty	0.67	0.63	**0.80**	0.71	-	-
LSTM [13]	0.60	0.63	0.60	0.61	0.64	0.70
ST-GCN [25]	0.63	0.61	0.83	0.70	0.69	0.64

CoT has higher precision but lower recall than certainty, suggesting CoT is more selective in classifying movements as correct. On the other hand, certainty has higher recall but lower precision than CoT, indicating more movements are classified as correct, which includes more false positives. This trade-off should be considered in the real-world implementation of this system, should healthcare professionals want to adjust sensitivity levels.

Despite their individual strengths, combining CoT with certainty did not yield improved results. Certainty-based prompting has similar effects to CoT reasoning in LLM classification tasks, as it implicitly prompts the model to justify its predictions. This suggests that certainty prompting slightly outperforms CoT by reducing CoT-induced "hallucinations" [36]. In CoT experiments, erroneous reasoning was found to lead to an accumulation of mistakes. Additionally, zero-shot CoT sometimes caused the LLM to create its own classification thresholds, which could either exaggerate errors or misalign with rehabilitation clinicians' expectations. For instance, the LLM outputted: "the shoulder abduction angle reaches a maximum of 160°, which is significantly higher than the expected 150°." This not only exaggerated the mistake but also imposed a threshold of 150°, whereas UI-PRMD physiotherapists may consider shoulder abductions non-optimal when the patient exhibits less than 160° of abduction. Consequently, the LLM would classify subsequent samples within the same conversation based on this initially self-assigned hard threshold. This trend was most pronounced when CoT reasoning was explicitly extracted. This highlights a key motivation for using LLMs rather than fixed threshold-based algorithms, as rehabilitation assessment requires adaptive interpretation of patient movement patterns, which may vary significantly across patients. Additionally, the reasoning-elicitation experiments reveal that likely explanations are not always correct, as most of the certainty and probability scores generated by the LLM ranged between 0.8 and 1, regardless of the actual accuracy of the assessment. This indicates that the broader issue of overconfidence in LLMs persists in the context of exercise quality assessment [35, 36].

To compare the performance of the LLM with previous deep learning techniques [13, 25], the last two rows of Table 4 (a) and (b) present the results of LSTM and ST-GCN on UI-PRMD and REHAB24-6. The LSTM model features a two-layer architecture, with each layer containing 64 hidden units, followed by a fully connected layer of size 64×1, where the single output dimension corresponds to the binary classification task. The ST-GCN model comprises three ST-GCN layers, followed by an average pooling layer, as designed in [25]. The output from the pooling layer is further processed by a convolutional layer that maps it to a single output dimension for classification. As shown in Table 4 (a) and (b), while the LLM, with any prompting technique, outperformed both LSTM and ST-GCN for REHAB24-6, its performance was inferior to that of LSTM and ST-GCN for UI-PRMD. The higher inter-class separability of data samples in UI-PRMD compared to REHAB24-6 reduces classification complexity, making it more suitable for traditional deep-learning models. Although the LLM performs worse than LSTM and ST-GCN on UI-PRMD, it is still advantageous by providing feedback and reasoning on correct and incorrect classifications, enhancing interpretability beyond conventional approaches.

4.3   Exercise-Specific Results Across Datasets

Building on the optimal prompt settings from previous experiments, three-shot prompting with extracted features and certainty elicitation was tested across common movements in REHAB24-6 and UI-PRMD.

Differences in performance across different movements were observed in Table 5, likely due to inherent variations in extracted features and their effectiveness in capturing movement errors. For both datasets, the squat exercise (ex6, m01) was particularly well-suited for LLM evaluation. The inherent symmetry of this movement likely creates a better distinction between correct and incorrect executions. In contrast, exercises such as leg lunges and shoulder abductions primarily involve the movement of a single body part, potentially leading to oversight of other contributing factors to incorrect movements. For instance, if a

Table 5. Performance comparison of GPT-4o on rehabilitation exercise quality assessment for the three common exercises in REHAB24-6 and UI-PRMD: shoulder abduction (ex1, m07), leg lunge (ex5, m03), and squat (ex6, m01), refer to Table 1.

	REHAB24-6			UI-PRMD		
Exercise	ex1	ex5	ex6	m07	m03	m01
Accuracy	0.67	0.74	**0.75**	**0.76**	0.67	**0.76**
Precision	0.71	0.69	**0.78**	**0.76**	0.64	0.69
Recall	0.75	**0.90**	0.70	0.84	0.76	**0.95**
F1	0.73	**0.78**	0.74	**0.80**	0.70	**0.80**

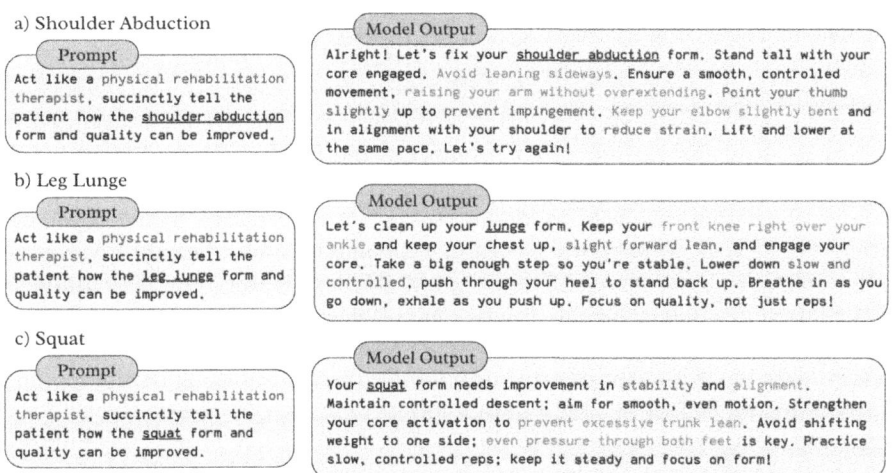

Fig. 3. Role-playing prompts for feedback generation following classification and reasoning, applied to (a) shoulder abduction, (b) leg lunge, and (c) squat exercise. The corresponding classification results are presented in Table 5. The LLM generates textual feedback in the specified role's style, incorporating suggestions derived from data trends (highlighted in green) and insights from its prior knowledge (highlighted in red).

patient performs a leg lunge but exhibits subtle instability in their upper body, which is not explicitly included in the analysis, the LLM may classify the movement as correct, whereas a rehabilitation expert would identify the movement as incorrect.

Feedback Evaluation. After obtaining the exercise quality prediction from the LLM, a role-play prompt was employed to generate concise textual feedback. Providing textual feedback after the model has already assessed exercise quality offers additional insight into its reasoning process. As illustrated in Fig. 3, prompting the LLM to explain how the patient can correct their movement by utilizing the role-playing technique within our two-step framework. This approach has been shown to enhance trust in initial responses and improve certainty estimations [41] by granting the LLM a second opportunity for reasoning and justification. Moreover, the LLM-generated feedback not only directly addresses the extracted features but also incorporates prior knowledge and general movement expectations. These include considerations such as stability, controlled movement speed, reducing strain, and breathing techniques, further enriching the quality of the feedback.

5   Conclusion and Future Works

This paper proposed a novel method utilizing LLMs for rehabilitation exercise quality assessment and feedback generation. By integrating advanced prompting techniques with exercise-specific features extracted from body joint sequences, the method enabled LLMs to analyze features, assess exercise quality, and generate meaningful feedback. The proposed approach achieved high accuracy in exercise assessments with few-shot prompting and produced valuable feedback. We also observed that the LLM remained overconfident of their outcomes, irrespective of whether the decision was correct or not. Our results demonstrate the potential of LLMs to support accurate, explainable, and adaptable AI-driven virtual rehabilitation systems. Despite promising results, the proposed approach has certain limitations. One key limitation is the lack of quantitative evaluation for feedback generation due to the absence of publicly available rehabilitation exercise datasets with ground-truth textual feedback. Additionally, the method relies on pre-trained LLMs, with data analysis performed based on their knowledge base. While prompting techniques such as few-shot prompting improve the LLMs' ability to evaluate exercises, fine-tuning LLMs specifically for this task could further enhance performance. Furthermore, the inability to set a random seed in the GPT-4o LLM affects result reproducibility. Since the GPT-4o LLM does not allow for controlled randomness, the generated feedback may vary across different runs, making it challenging to ensure consistent evaluation and comparison across different experiments. Future work will focus on collecting and annotating rehabilitation exercise datasets with ground-truth feedback and leveraging this data to fine-tune pre-trained LLMs, enhancing their performance and reliability in exercise quality assessment and feedback generation.

Funding. This research was funded by the New Frontiers in Research Fund, Canada, and the TRANSFORM HF Undergraduate Summer Research Program, Canada.

References

1. World Health Organization, "Rehabilitation," https://www.who.int/news-room/fact-sheets/detail/rehabilitation (2023). Accessed 30 Jan 2023
2. Dibben, G.O., et al.: Exercise-based cardiac rehabilitation for coronary heart disease: a meta-analysis. Eur. Heart J. **44**(6), 452–469 (2023)

3. Shirozhan, S., Arsalani, N., Seyed Bagher Maddah, S., Mohammadi-Shahboulaghi, F.: Barriers and facilitators of rehabilitation nursing care for patients with disability in the rehabilitation hospital: a qualitative study. Front. Public Health **10** (2022). https://doi.org/10.3389/fpubh.2022.931287

4. Ferreira, R., Santos, R., Sousa, A.: Usage of auxiliary systems and artificial intelligence in home-based rehabilitation: a review. In: Queirós, R., Cunha, B., Fonseca, X. (eds.) Exploring the Convergence of Computer and Medical Science Through Cloud Healthcare, pp. 163–196. IGI Global (2022). https://doi.org/10.4018/978-1-6684-5260-8.ch008

5. Seron, P., et al.: Effectiveness of telerehabilitation in physical therapy: a rapid overview. Phys. Ther. **101**(6) (2021). https://doi.org/10.1093/ptj/pzab053

6. Boukhennoufa, I., Zhai, X., Utti, V., Jackson, J., McDonald-Maier, K.D.: Wearable sensors and machine learning in post-stroke rehabilitation assessment: a systematic review. Biomed. Signal Process. Control **71**, 103197 (2022)

7. Abedi, A., Colella, T.J., Pakosh, M., Khan, S.S.: Artificial intelligence-driven virtual rehabilitation for people living in the community: a scoping review. NPJ Digital Med. **7**(1), 25 (2024)

8. Sardari, S., et al.: Artificial intelligence for skeleton-based physical rehabilitation action evaluation: a systematic review. Comput. Biol. Med. **158**, 106835 (2023). https://doi.org/10.1016/j.compbiomed.2023.106835

9. Ettefagh, A., Roshan Fekr, A.: Technological advances in lower-limb telerehabilitation: a review of literature. J. Rehab. Assist. Technol. Eng. **11** (2024). https://doi.org/10.1177/20556683241259256

10. Brennan, L., Dorronzoro Zubiete, E., Caulfield, B.: Feedback design in targeted exercise digital biofeedback systems for home rehabilitation: a scoping review. Sensors **20**(1), 181 (2019). https://doi.org/10.3390/s20010181

11. Pavllo, D., Feichtenhofer, C., Grangier, D., Auli, M.: 3D human pose estimation in video with temporal convolutions and semi-supervised training. In: Proceedings of the IEEE/CVF Conference on Computer Vision and Pattern Recognition, pp. 7753–7762 (2019)

12. Lugaresi, C.: Mediapipe: A framework for building perception pipelines. arXiv preprint arXiv:1906.08172 (2019)

13. Abedi, A., Malmirian, M., Khan, S.: Cross-modal video to body-joints augmentation for rehabilitation exercise quality assessment. arXiv preprint arXiv:2306.09546 (2023)

14. Yan, S., Xiong, Y., Lin, D.: Spatial temporal graph convolutional networks for skeleton-based action recognition. Proc. AAAI Conf. Artif. Intell. **32**(1) (2018). https://doi.org/10.1609/aaai.v32i1.12328

15. Capecci, M., et al.: The Kimore dataset: Kinematic assessment of movement and clinical scores for remote monitoring of physical rehabilitation. IEEE Trans. Neural Syst. Rehabil. Eng. **27**(7), 1436–1448 (2019). epub 2019 Jun 14

16. Schuber, A., Schaller, A.: Relevance of therapist feedback in the context of group-based exercise programs in medical rehabilitation–results from a qualitative study with patients and exercise therapists. Europ. J. Physiother. 1–9 (2024)

17. Popović, M.D., Kostić, M.D., Rodić, S.Z., Konstantinović, L.M.: Feedback-mediated upper extremities exercise: Increasing patient motivation in poststroke rehabilitation. Biomed. Res. Int. **2014**(1), 520374 (2014)

18. Wang, C., et al.: Ubiphysio: support daily functioning, fitness, and rehabilitation with action understanding and feedback in natural language. Proc. ACM Interact., Mobile, Wearable Ubiquitous Technol. **8**(1), 1–27 (2024)

19. Vakanski, A., Jun, H., Paul, D., Baker, R.: A data set of human body movements for physical rehabilitation exercises. Data **3**(1), 2 (2018). https://doi.org/10.3390/data3010002

20. Černek, A., Sedmidubsky, J., Budikova, P.: REHAB24-6: physical therapy dataset for analyzing pose estimation methods. In: Chávez, E., Kimia, B., Lokoč, J., Patella, M., Sedmidubsky, J. (eds.) Similarity Search and Applications: 17th International Conference, SISAP 2024, Providence, RI, USA, November 4–6, 2024, Proceedings, pp. 18–33. Springer Nature Switzerland, Cham (2025). https://doi.org/10.1007/978-3-031-75823-2_2

21. Liao, Y., Vakanski, A., Xian, M.: A deep learning framework for assessing physical rehabilitation exercises. IEEE Trans. Neural Syst. Rehabil. Eng. **28**(2), 468–477 (2020)

22. Guo, Q., Khan, S.S.: Exercise-specific feature extraction approach for assessing physical rehabilitation. In: 4th IJCAI Workshop on AI for Aging. Rehabilitation and Intelligent Assisted Living, IJCAI (2021)

23. Karagoz, B., Ashraf, A., Khan, S.: Supervised sequential contrastive regression: Improving performance on imbalanced rehabilitation exercises datasets. preprint, 12 (2023)

24. Deb, S., Islam, M.F., Rahman, S., Rahman, S.: Graph convolutional networks for assessment of physical rehabilitation exercises. IEEE Trans. Neural Syst. Rehabil. Eng. **30**, 410–419 (2022)

25. Zheng, K., Wu, J., Zhang, J., Guo, C.: A skeleton-based rehabilitation exercise assessment system with rotation invariance. IEEE Trans. Neural Syst. Rehab. Eng. **31**, 2612–2621 (2023). https://doi.org/10.1109/TNSRE.2023.3282675

26. Réby, K., Dulau, I., Dubrasquet, G., Aimar, M.B.: Graph transformer for physical rehabilitation evaluation. In: 2023 IEEE 17th International Conference on Automatic Face and Gesture Recognition (FG). IEEE, 2023, pp. 1–8 (2023)

27. Karlov, M., Abedi, A., Khan, S.S.: Rehabilitation exercise quality assessment through supervised contrastive learning with hard and soft negatives. Med. Biol. Eng. Comput. **63**(1), 15–28 (2025). https://doi.org/10.1007/s11517-024-03177-x

28. Bruce, X., Liu, Y., Chan, K.C., Chen, C.W.: EGCN++: A new fusion strategy for ensemble learning in skeleton-based rehabilitation exercise assessment. IEEE Trans. Pattern Anal. Mach. Intell. (2024)

29. Parker, J., Mountain, G., Hammerton, J.: A review of the evidence underpinning the use of visual and auditory feedback for computer technology in post-stroke upper-limb rehabilitation. Disabil. Rehabil. Assist. Technol. **6**(6), 465–472 (2011)

30. Li, J.: Finerehab: A multi-modality and multi-task dataset for rehabilitation analysis. In: Proceedings of the IEEE/CVF Conference on Computer Vision and Pattern Recognition, pp. 3184–3193 (2024)

31. Miron, A., Sadawi, N., Ismail, W., Hussain, H., Grosan, C.: IntelliRehabDS (IRDS)—a dataset of physical rehabilitation movements. Data **6**(5), 46 (2021). https://doi.org/10.3390/data6050046

32. Zaghir, J., Naguib, M., Bjelogrlic, M., Névéol, A., Tannier, X., Lovis, C.: Prompt engineering paradigms for medical applications: Scoping review. J. Med. Internet Res. **26**, e60501 (2024)

33. Brown, T., et al.: Language models are few-shot learners. Adv. Neural. Inf. Process. Syst. **33**, 1877–1901 (2020)

34. Touvron, H., et al.: Llama: Open and efficient foundation language models. arXiv preprint arXiv:2302.13971 (2023)

35. Xiong, M., Hu, Z., Lu, X., Li, Y., Fu, J., He, J., Hooi, B.: Can LLMs express their uncertainty? an empirical evaluation of confidence elicitation in LLMs. arXiv preprint arXiv:2306.13063 (2023)
36. Becker, E., Soatto, S.: Cycles of thought: Measuring LLM confidence through stable explanations. arXiv preprint arXiv:2406.03441 (2024)
37. Wei, J., et al.: Chain-of-thought prompting elicits reasoning in large language models. In: Advances in Neural Information Processing Systems, vol. 35, pp. 24 824–24 837 (2022)
38. Gu, B., Desai, R.J., Lin, K.J., Yang, J.: Probabilistic medical predictions of large language models. NPJ Digital Med. **7**(1) (2024). https://doi.org/10.1038/s41746-024-01366-4
39. Shanahan, M., McDonell, K., Reynolds, L.: Role play with large language models. Nature **623**(7987), 493–498 (2023)
40. OpenAI, "Gpt-4o announcement," (2024). Accessed 17 Feb 2025. https://openai.com/index/hello-gpt-4o
41. Li, M., Wang, W., Feng, F., Zhu, F., Wang, Q., Chua, T.-S.: Think twice before assure: Confidence estimation for large language models through reflection on multiple answers. arXiv preprint arXiv:2403.09972 (2024)

Transforming Digital Reminder Systems for Dementia Care into Behavioral Anomaly Detectors: A Proof-of-Concept Using LSTM Autoencoders

Joy Lai$^{(\boxtimes)}$ ⓘ and Alex Mihailidis ⓘ

Institute of Biomedical Engineering, University of Toronto, Toronto, Canada
joy.lai@mail.utoronto.ca, alex.mihailidis@utoronto.ca

Abstract. Digital reminder systems are widely used to support daily routines among older adults, especially people living with dementia (PLwD). While effective at issuing prompts, these systems typically lack mechanisms to monitor user engagement or detect subtle behavioral changes that may indicate early signs of dementia progression or need for caregiver intervention. This study presents a proof-of-concept for transforming digital reminder systems into behavioral anomaly detectors using LSTM (Long Short-Term Memory) autoencoders. Through simulated datasets generated from smart home interaction logs, we model two caregiver-prioritized anomalies—delayed acknowledgments and location-based ignoring of reminders. We evaluate the performance of vanilla LSTM autoencoders, classifier-augmented variants, and traditional methods such as Z-score thresholding and Isolation Forest. Our results show that the LSTM autoencoder combined with a Random Forest classifier consistently outperforms all other models across all anomaly types and severity levels, achieving the highest detection performance in every condition tested. This work demonstrates the feasibility of using existing reminder systems as privacy-preserving, low-friction platforms for early behavioral monitoring in dementia care, with implications for enhancing caregiver support and clinical insight without requiring additional hardware or invasive sensing.

Keywords: Dementia Care · Anomaly Detection · Digital Reminders · LSTM Autoencoder · Behavioral Monitoring

1 Introduction

Managing daily routines becomes increasingly difficult for older adults experiencing cognitive decline, particularly people living with dementia (PLwD) [1]. Symptoms such as memory loss, confusion, and reduced attention span can disrupt essential tasks like taking medication, maintaining hygiene, and attending appointments [2, 3]. To support independence and reduce caregiver stress, digital reminder systems—delivered via smartphones, tablets, or smart home devices—have been widely adopted [4].

© The Author(s), under exclusive license to Springer Nature Singapore Pte Ltd. 2025
S. S. Khan et al. (Eds.): IJCAI 2025, CCIS 2620, pp. 76–90, 2025.
https://doi.org/10.1007/978-981-95-0568-5_6

However, these systems are largely passive: they issue prompts but rarely monitor how users engage with them over time [4–7]. While missed reminders can sometimes be logged, more nuanced behavioral shifts—like increasing response delays or declining usage—often go unnoticed [6–8]. These patterns may reflect early signs of dementia progression or indicate a need for intervention, but current systems generally lack the mechanisms to detect such trends.

Although reminder interactions are rich in behavioral data, anomaly detection has rarely been applied in this domain [9]. Most healthcare-related efforts have focused on ambient or physiological sensors [9, 10]. In this paper, we explore whether reminder systems themselves—already integrated into daily life—can serve as a novel, underused stream for behavioral anomaly detection.

As a proof of concept, we assess whether LSTM (Long Short-Term Memory) autoencoders can model and detect deviations in interactions with the reminder system through reminder metadata. These models capture temporal dependencies and flag anomalies through reconstruction error [11–15]. We compare their performance to traditional methods like Isolation Forest and Z-score thresholding. This work was developed with input from two dementia caregivers through a collaboration with the Engagement of People with Lived Experience of Dementia (EPLED), research advisory group comprised of PLwD and caregivers [16]. They identified two high-priority behavioral anomalies: 1) Delays in reminder acknowledgment time, and 2) Ignoring reminders in specific locations. These behaviors may indicate early disengagement or context-sensitive lapses that warrant attention. Accordingly, this study focuses on these two anomaly types using synthetically augmented data. This study's contributions are as follows:

- A simulation framework for generating structured reminder interaction data with task and location context based on real smart home logs.
- Injection of behaviorally grounded anomalies—delayed acknowledgments and location-based ignoring—prioritized through caregiver input.
- Application and benchmarking of LSTM autoencoders, with and without classifier augmentation, against traditional anomaly detection methods.
- Demonstration that classifier-augmented models significantly improve detection performance.
- Evidence that reminder systems can be repurposed as accessible, privacy-preserving tools for early cognitive monitoring.

2   Related Work

2.1   Reminder and Task Management Systems

Digital reminder systems are widely used assistive technologies that help older adults and PLwD manage daily routines [4]. These range from basic calendar alerts to advanced multimodal platforms delivering prompts through smartphones, tablets, or smart home devices. In dementia care, systems like eSticky and MindMate have shown success in promoting routine adherence and reducing the burden of repetitive prompting [4–7].

Despite their utility, most systems remain unidirectional—delivering prompts without interpreting user engagement over time [4]. While some may log missed reminders, they often overlook subtle patterns such as delayed responses, selective ignoring, or reduced usage [4–7]. Current tools rely on caregivers to detect such changes, limiting their potential for proactive support. Moreover, by automating prompts without automating monitoring, these systems may inadvertently increase caregiver workload [17, 18]. Caregivers must still verify task completion or follow up manually, reflecting a broader issue in assistive tech design: task automation does not always equate to reduced oversight workload [19, 20].

To address these limitations, we are developing a location-based reminder system in which prompts are delivered via tablets placed in key household areas [8]. For instance, a kitchen tablet might display a reminder to turn off the stove, while a front-door unit could prompt a user to grab their keys. This setup provides context-sensitive prompts and generates rich interaction data. Building on this foundation, we propose a novel extension: a system that can interpret engagement patterns and detect behavioral anomalies. Such a system could reduce manual monitoring demands while offering timely insights into user adherence.

2.2 Anomaly Detection in Healthcare and Activities of Daily Living

Anomaly detection is an important approach in aging and dementia care, particularly for monitoring activities of daily living in home settings [9]. Deviations such as skipped meals, hygiene lapses, or disrupted sleep are often inferred from smart home sensors, motion detectors, or wearable devices [21–24]. These behavioral shifts can signal emerging risks, including cognitive decline, functional dependence, or safety concerns [22, 25, 26].

However, adoption of these technologies faces practical barriers. Wearables require consistent use and can be stigmatizing or uncomfortable [27]. In-home sensors and cameras raise privacy concerns and may be expensive or difficult to install—especially in rental units or low-resource environments [18, 28]. These limitations hinder the scalability of many current solutions.

In contrast, digital reminder systems are already integrated into the daily lives of many PLwD and their caregivers [4]. They typically run on familiar devices like smartphones, tablets, or smart displays and require no additional hardware. This makes them a promising, privacy-preserving alternative for behavioral monitoring. If interaction patterns with reminders can be analyzed for signs of cognitive decline, such systems could offer a low-cost, low-friction path to early detection.

Despite this potential, anomaly detection using reminder system data remains underexplored. In this work, we focus on the previously described location-based reminder system that delivers context-aware prompts via tablets placed throughout the home [8]. We investigate whether this system can function not only as a task prompter but also as a passive behavioral sensor capable of identifying early signs of disorientation or disengagement.

2.3 LSTM Autoencoders for Time-Series Anomaly Detection

LSTM autoencoders are a common method for unsupervised anomaly detection in sequential data [12, 13]. These models learn typical temporal patterns and reconstruct input sequences; elevated reconstruction error can indicate anomalous behavior [11, 14,

15]. Their ability to capture long-range dependencies and irregular time series makes them especially suited for modeling complex human behaviors. In healthcare, LSTM autoencoders have been used to detect anomalies in ECG signals, glucose trends, and daily activity patterns [14, 29, 30]. They've also proven effective in domains such as industrial fault detection, financial forecasting, and cybersecurity [12, 13, 15, 31].

In this work, we apply LSTM autoencoders to a novel domain: digital reminder interactions. Unlike ambient sensor data, reminder events are intentional—each prompt anticipates a response. Deviations in this context (e.g., delayed acknowledgments or missed responses) may offer meaningful indicators of behavioral anomalies. To our knowledge, this is the first application of LSTM autoencoders to this data type. Key advantages of LSTM autoencoders for this task include:

- Modeling sequential context to detect subtle behavioral shifts;
- Unsupervised learning, well-suited to domains with scarce labeled anomalies;
- Continuous anomaly scoring, enabling real-time or classifier-augmented analysis.

2.4 Summary of Research Gap

Although deep learning models—particularly autoencoders—have shown promise in detecting behavioral anomalies among aging populations, their use has largely focused on passive sensing technologies such as PIR motion detectors, RFID tags, cameras, and temperature sensors [32–34]. These methods often raise concerns about privacy, cost, installation complexity, and user acceptance [17, 35]. By contrast, digital reminder systems are already widely used and generate structured behavioral data. Yet they remain underutilized for anomaly detection due to their lack of interpretive feedback mechanisms [4]. Enhancing these systems with anomaly detection capabilities could offer a low-cost, privacy-conscious alternative—especially for users and caregivers already relying on such platforms.

This paper addresses that gap by evaluating whether LSTM autoencoders can detect meaningful deviations in reminder interactions. We use a simulation-based feasibility approach grounded in the priorities of dementia caregivers. Specifically, we focus on two caregiver-identified anomaly types: delayed acknowledgments and location-based ignoring. By applying sequence modeling to this structured yet overlooked data stream, we explore the potential to transform reminder systems into accessible behavioral monitoring tools.

3 Methodology

3.1 Dataset

Dataset Preparation and Preprocessing
We used 31 CASAS smart home datasets, which include annotated sensor data capturing residents' daily activities, to generate synthetic reminder interaction sequences [36]. Each dataset was transformed to simulate structured digital reminders, including attributes such as type, location, display time, acknowledgment status and time,

and priority. Reminder types included mealtime routines, hygiene, medical appointments, and social engagement. Acknowledgment behavior was probabilistically modeled based on context, such as reminder type and location. Datasets were partitioned by size: smaller datasets (<400 rows) were allocated entirely to training, medium datasets (400–1000 rows) were split 70/30 into training and testing, and larger datasets (>1000 rows) were reserved for testing. Preprocessing included embedding of categorical fields and normalization of relevant features for time-series modeling.

Anomaly Injection Strategy

To enable controlled evaluation, we injected synthetic anomalies into test data. We defined three types of behaviorally grounded anomalies:

1. Increased acknowledgment delay
2. Location-based ignoring
3. Combined behavior shift (delay + ignoring)

Each type was implemented at three severity levels (mild, moderate, severe), as summarized in Table 1. Delay anomalies increased acknowledgment time by a percentage relative to the original delay. Location-based ignoring involved probabilistically converting acknowledged reminders from a selected location into non-responses; as a result, these anomalies affected fewer rows overall, since they were limited to reminders issued from one location. Combined anomalies applied both delay and location effects simultaneously. Datasets retained their original structure with an added anomaly label for evaluation and classifier training.

3.2 Model Training

We evaluated three approaches to anomaly detection:
A **vanilla LSTM autoencoder**, which identifies anomalies based on reconstruction error thresholds.

Two **classifier-augmented LSTM autoencoders**, which uses latent features and reconstruction errors to train supervised classifiers (random forest and logistic regression).

Two **baseline methods**—Z-score thresholding and Isolation Forest—for comparison.

Baseline and Sequence-Based Models

The vanilla LSTM autoencoder was trained on clean reminder interaction sequences using mean squared error as the loss function. Anomalies were flagged when reconstruction errors exceeded the 95th percentile of validation scores. For baselines, the Z-score method flagged sequences with any feature exceeding three standard deviations from the mean. Isolation Forest, an unsupervised algorithm, detected anomalies by identifying points that required fewer splits in a randomized decision tree ensemble—indicating their deviation from the norm.

Classifier-Augmented LSTM Autoencoder

To improve precision and sensitivity, particularly for subtle anomalies, we combined the

Table 1. Overview of the Anomaly Injection Strategy

Anomaly Type	Description	Affected Features	Severity Levels
Increased Acknowledgment Delay	Gradual increases in acknowledgment time for reminders that were originally acknowledged	Acknowledgment Time	**Mild**: + 10–30% delay **Moderate**: + 30–60% delay **Severe**: + 50–100% delay
Location-Based Ignoring	Selective non-responsiveness to reminders triggered from specific locations	Acknowledgment Status Acknowledgment Time Location	**Mild**: 10–30% reminders ignored **Moderate**: 30–60% ignored **Severe**: 50–100% ignored
Combined Behavior Shift	Co-occurrence of both increased delay and selective ignoring, simulating complex disengagement	Acknowledgment Time/Status, Location	**Mild**: + 10–30% delay + partial ignore **Moderate**: + 30–60% delay + ~ 50% ignore **Severe**: + 50–100% delay + near-total ignore

LSTM autoencoder with supervised classifiers that detect anomalies using latent representations and reconstruction statistics—rather than relying on a fixed reconstruction error threshold. After training the autoencoder, synthetic anomalies were injected into a copy of the training data. Sequences were passed through the model to extract latent vectors and reconstruction statistics. These features were labeled and used to train two classifiers: logistic regression and random forest. During inference, test sequences were encoded in the same way and classified as anomalous or not, enabling more nuanced detection than thresholding alone.

3.3 Evaluation Metrics

We evaluated model performance using four standard metrics: **precision**, **recall**, **F1 score**, and **ROC-AUC**.

- **Precision** measures the proportion of predicted anomalies that are correct. High precision minimizes false positives—crucial when flagging potential cognitive risks for caregiver attention.
- **Recall** (sensitivity) captures the proportion of actual anomalies correctly identified. High recall is essential for detecting subtle behavioral changes.
- **F1 Score** is the harmonic mean of precision and recall, providing a balanced measure when both false positives and false negatives matter.
- **ROC-AUC** (Area Under the Receiver Operating Characteristic Curve) assesses the model's ability to distinguish anomalies from normal sequences across all thresholds.

A higher value reflects better overall discriminative power, independent of specific cutoffs.

Metrics were calculated for each anomaly type and severity level to enable detailed comparison across models and detection strategies, with results averaged across all test datasets.

4 Results

Across all anomaly types and severity levels, the LSTM autoencoder combined with a Random Forest classifier consistently outperformed other models. As shown in Table 2, it achieved F1-scores near or above 0.99, with precision and recall values typically exceeding 0.97 for the acknowledgement delay and combined anomalies. Even on the more challenging anomaly, location-based ignoring, the model still performed significantly better than all other approaches. A common trend across all models and anomaly types was that performance improved as anomaly severity increased—likely because more extreme deviations are easier to distinguish from normal behavior, making them more detectable even with simpler detection strategies.

In contrast, the vanilla LSTM autoencoder, while achieving high recall, suffered from lower precision, leading to reduced F1-scores. This underscores the limitations of relying solely on reconstruction error thresholding for context-sensitive anomalies. The classifier-augmented approach significantly improved both sensitivity and specificity. Logistic Regression provided moderate gains over the vanilla model, particularly in improving precision. However, it struggled on the location-based ignoring anomalies, with F1-scores often remaining below 0.60.

Baseline methods, including Z-score thresholding and Isolation Forest, demonstrated clear limitations. Z-score handled acknowledgement delay and combined anomalies reasonably well but failed on spatial anomalies due to low precision. Isolation Forest generally underperformed across all types, rarely exceeding an F1-score of 0.45.

These results highlight the strength of sequence-based models, particularly when combined with supervised classifiers. The LSTM + Random Forest approach effectively captures subtle behavioral deviations that traditional baselines cannot.

Table 2. Full Metrics by Model, Anomaly Type, and Severity

Algorithm	Precision	Recall	F1 Score	ROC AUC
Increased Acknowledgment Delay, Mild				
Vanilla LSTM AE	0.556	0.904	0.658	0.528
LSTM AE + Logistic Regression	0.904	0.617	0.689	0.757
LSTM AE + Random Forest	0.998	0.975	0.987	0.992
Isolation Forest	0.511	0.420	0.419	0.474
Z-Score	0.556	0.961	0.688	0.486
Increased Acknowledgment Delay, Moderate				
Vanilla LSTM AE	0.521	0.889	0.629	0.503
LSTM AE + Logistic Regression	0.919	0.635	0.696	0.747
LSTM AE + Random Forest	0.997	0.995	0.996	0.996
Isolation Forest	0.508	0.468	0.452	0.502
Z-Score	0.537	0.964	0.676	0.507
Increased Acknowledgment Delay, Severe				
Vanilla LSTM AE	0.547	0.890	0.645	0.508
LSTM AE + Logistic Regression	0.948	0.652	0.711	0.845
LSTM AE + Random Forest	0.997	0.997	0.997	0.997
Isolation Forest	0.564	0.524	0.498	0.521
Z-Score	0.560	0.958	0.690	0.527
Location-Based Ignoring, Mild				
Vanilla LSTM AE	0.012	0.801	0.024	0.567
LSTM AE + Logistic Regression	0.422	0.496	0.387	0.535
LSTM AE + Random Forest	0.765	0.858	0.798	0.965
Isolation Forest	0.014	0.474	0.027	0.543
Z-Score	0.012	0.810	0.023	0.465
Location-Based Ignoring, Moderate				
Vanilla LSTM AE	0.024	0.823	0.046	0.509
LSTM AE + Logistic Regression	0.573	0.505	0.497	0.515
LSTM AE + Random Forest	0.862	0.904	0.880	0.962
Isolation Forest	0.025	0.350	0.044	0.520
Z-Score	0.023	0.827	0.045	0.522
Location-Based Ignoring, Severe				
Vanilla LSTM AE	0.064	0.861	0.118	0.490

(*continued*)

Table 2. (*continued*)

Algorithm	Precision	Recall	F1 Score	ROC AUC
LSTM AE + Logistic Regression	0.878	0.524	0.597	0.573
LSTM AE + Random Forest	0.914	0.906	0.910	0.967
Isolation Forest	0.074	0.415	0.117	0.567
Z-Score	0.065	0.875	0.117	0.493
Combined Behavior Shift, Mild				
Vanilla LSTM AE	0.529	0.901	0.636	0.524
LSTM AE + Logistic Regression	0.942	0.597	0.667	0.715
LSTM AE + Random Forest	0.997	0.988	0.992	0.994
Isolation Forest	0.494	0.427	0.419	0.479
Z-Score	0.540	0.967	0.680	0.499
Combined Behavior Shift, Moderate				
Vanilla LSTM AE	0.568	0.904	0.668	0.521
LSTM AE + Logistic Regression	0.988	0.596	0.689	0.770
LSTM AE + Random Forest	0.999	0.996	0.998	0.999
Isolation Forest	0.547	0.489	0.481	0.487
Z-Score	0.576	0.964	0.709	0.501
Combined Behavior Shift, Severe				
Vanilla LSTM AE	0.494	0.897	0.631	0.547
LSTM AE + Logistic Regression	0.975	0.599	0.686	0.755
LSTM AE + Random Forest	0.998	0.992	0.995	0.997
Isolation Forest	0.548	0.524	0.491	0.541
Z-Score	0.533	0.955	0.671	0.541

4.1 Analysis by Anomaly Type

As shown in Figs. 1, 2 and 3, model performance varied across anomaly types, with the LSTM AE + Random Forest consistently achieving the highest F1 and ROC-AUC scores across all conditions. The following section provides a breakdown of results for each anomaly type.

Increased Acknowledgment Delay

This was the simplest anomaly to detect due to its clear, systematic deviation in response time. As shown in Fig. 1, most models performed reasonably well, but the LSTM AE + Random Forest consistently outperformed all others, with near-perfect F1 and ROC-AUC scores. Simpler models like Z-score and Logistic Regression showed moderate success, indicating that delay-based anomalies are detectable even with basic features. The vanilla LSTM AE had high recall but lower precision, while Isolation Forest performed poorly across all metrics.

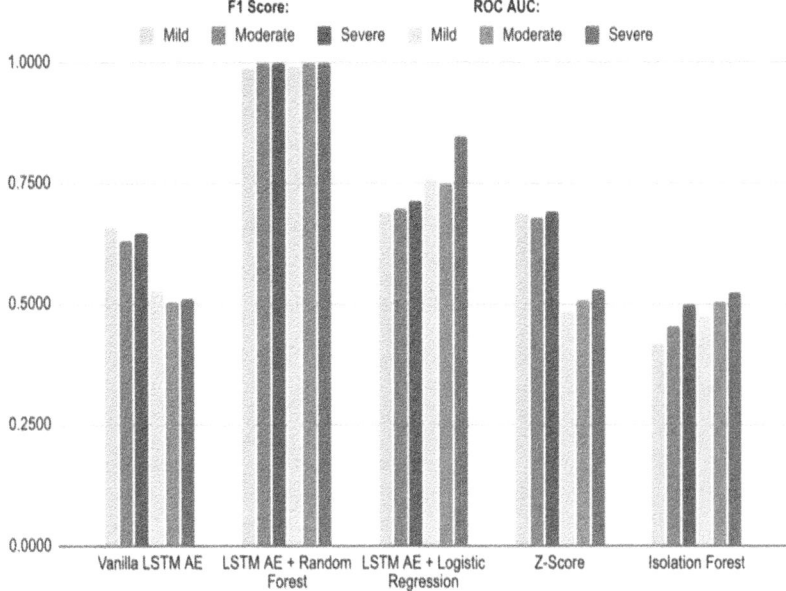

Fig. 1. F1 Scores (purple) and ROC AUC (blue) for "Increased Acknowledgment Delay" across detection models and severity levels

Location-Based Ignoring

This type proved more difficult to detect because it affected only a subset of reminders—those triggered from a single location—resulting in fewer impacted rows within each dataset. Additionally, the behavioral deviations were often gradual and subtle, making them harder to distinguish from normal variation. As shown in Fig. 2, while recall was often high for unsupervised models, precision was extremely low, resulting in poor F1-scores—especially for the vanilla AE, Z-score, and Isolation Forest. Logistic Regression offered slight improvements, but only the LSTM AE + Random Forest maintained robust performance across all severities, suggesting its ability to capture nuanced engagement patterns that simpler methods miss.

Combined Behavior Shift

Combining delayed responses and location-based ignoring led to stronger performance across most models. Again, as shown in Fig. 3, the LSTM AE + Random Forest achieved near-perfect scores. The additive nature of these anomalies helped boost statistical detectability, improving Z-score and Logistic Regression performance somewhat—but precision remained a limiting factor. Isolation Forest continued to underperform.

5   Discussion

This study examined whether digital reminder systems—typically passive tools—could also serve as behavioral data sources for anomaly detection. Using LSTM autoencoders, we explored whether patterns like delayed responses or location-specific

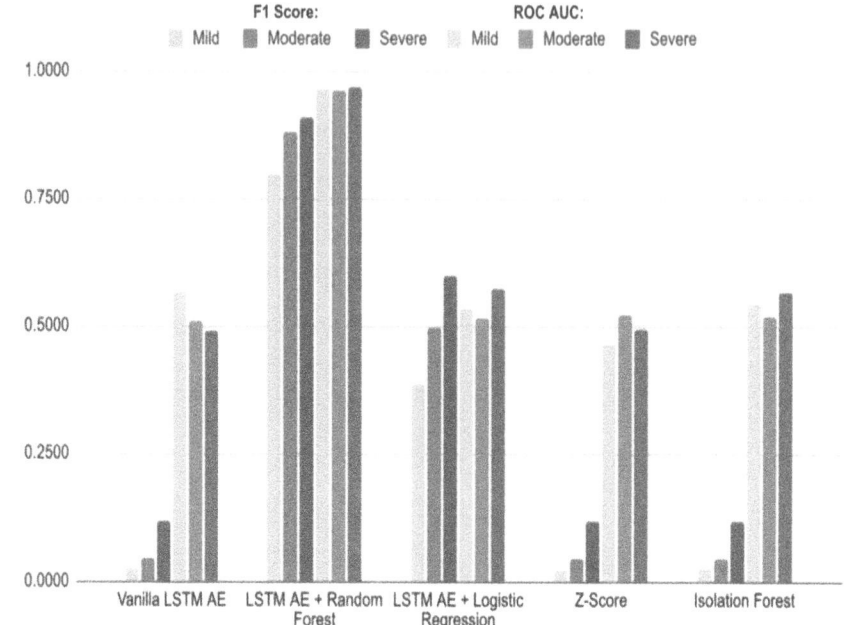

Fig. 2. F1 Scores (purple) and ROC AUC (blue) for "Location-Based Ignoring" across detection models and severity levels

non-responsiveness could be detected consistently. As a feasibility study, this work applies sequence-based modeling to a domain where behavioral labels are scarce but semantically meaningful.

Our findings show that temporal modeling, especially when paired with supervised classifiers, significantly outperforms traditional anomaly detection methods. While

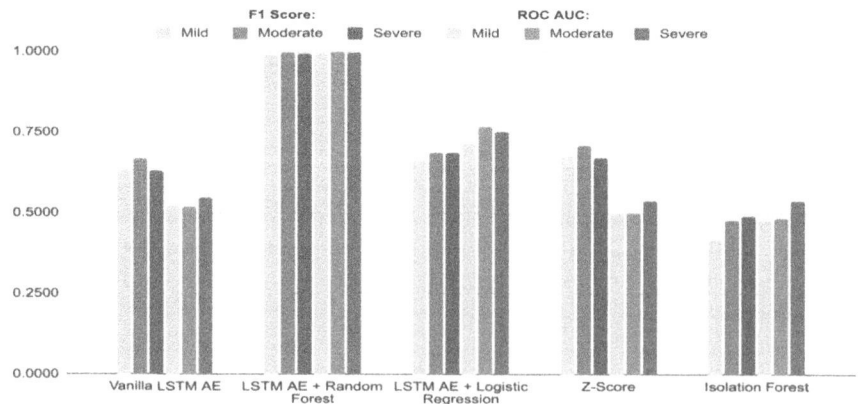

Fig. 3. F1 Scores (purple) and ROC AUC (blue) for "Combined Behavior Shift" across detection models and severity levels

the vanilla LSTM autoencoder effectively captured anomalies with strong temporal signals, its precision dropped on context-dependent patterns like location-based ignoring—highlighting a common limitation of threshold-based methods.

Adding lightweight classifiers—especially Random Forest—to the autoencoder improved precision and overall accuracy across all anomaly types. Even simpler classifiers like logistic regression provided benefit, showing that once robust sequence features are learned, supervised models can better detect subtle behavioral shifts.

In contrast, baseline methods such as Isolation Forest and Z-score thresholding performed inconsistently. While capable of detecting large deviations, they lacked the precision needed for real-world deployment—especially with spatial or contextual anomalies. Overall, these results underscore the value of sequence-aware, context-sensitive modeling in turning passive reminder interactions into actionable behavioral insights.

5.1 Implications for Dementia Care

Beyond technical contributions, this work suggests a new role for digital reminder systems: passive monitoring of behavioral consistency. In dementia care, where subtle engagement changes may precede clinical symptoms, detecting such deviations early could support more personalized interventions, adaptive task scheduling, and timely caregiver alerts—without relying on wearables or cameras.

By automatically surfacing meaningful behavioral shifts, these systems could reduce caregivers' need to monitor routines manually. Additionally, they could generate a structured behavioral record over time, offering clinicians objective insights into at-home task adherence. This may ease caregiver documentation workload and support more informed clinical decisions.

5.2 Limitations and Future Work

While promising, this study has several limitations. All anomalies were synthetically generated; although these were grounded in caregiver-identified priorities, real-world validation in clinical or residential settings is necessary. Additionally, the models were trained on pooled datasets without individual personalization, which may limit their ability to adapt to the natural variability in behavior across users.

Future work should aim to develop classifiers that can distinguish between specific types of anomalies—such as delayed acknowledgment versus location-based ignoring—to provide more actionable feedback to caregivers. Further engagement with caregivers will be essential to identify additional behaviors they consider meaningful for monitoring, helping expand the system's relevance and usability. It will also be important to explore the best methods for surfacing anomalies: when alerts should be triggered, how they should be framed, and what contextual information caregivers need to interpret them effectively. Another key direction involves evaluating whether detected anomalies reliably correspond to meaningful outcomes, such as cognitive decline, disengagement, or evolving care needs. Finally, building models that continuously learn from longitudinal behavioral data could improve personalization, reduce false positives, and enhance the system's long-term support for PLwD.

6 Conclusion

This study presents a novel application of LSTM autoencoders for detecting behavioral anomalies in digital reminder system interactions. Even the vanilla LSTM autoencoder-effectively learned engagement patterns and flagged deviations associated with cognitive or functional concerns. When combined with lightweight classifiers, detection performance improved substantially—especially for subtle, context-sensitive anomalies. Classical baselines lacking temporal or contextual modeling consistently underperformed. These findings suggest that digital reminder systems, already widely adopted in dementia care, could evolve into intelligent, feedback-aware tools. Rather than relying solely on caregiver oversight, they could autonomously surface early risk indicators, enhancing support for independent living. In doing so, such systems offer an alternative means of behavioral monitoring for caregivers—reducing the workload of manual observation while still ensuring timely insights. Embedding anomaly detection into everyday assistive technologies offers a practical path toward proactive, behavior-aware cognitive support.

Acknowledgments. This study was supported by AGE-WELL NCE and the University of Toronto. The authors thank Kelly Beaton and David Black for their valuable and ongoing contributions as lived experience experts.

Disclosure of Interests. The authors have no competing interests to declare that are relevant to the content of this article.

References

1. Cipriani, G., Danti, S., Picchi, L., Nuti, A., Di Fiorino, M.: Daily functioning and dementia. Dementia e Neuropsychologia. **14**, 93–102 (2020). https://doi.org/10.1590/1980-57642020dn14-020001
2. Norton, L.E., Malloy, P.F., Salloway, S.: The impact of behavioral symptoms on activities of daily living in patients with Dementia. Am. J. Geriatr. Psychiatry **9**, 41–48 (2001)
3. Curnow, E., Rush, R., Maciver, D., Górska, S., Forsyth, K.: Exploring the needs of people with dementia living at home reported by people with dementia and informal caregivers: a systematic review and Meta-analysis. Aging Ment. Health **25**, 397–407 (2021). https://doi.org/10.1080/13607863.2019.1695741
4. Peres, B., Campos, P.F.: A systematic review of reminder and guidance systems for Alzheimer's Disease and Related Dementias patients: context, barriers and facilitators. Disabil. Rehabil. Assist. Technol. (2023). https://doi.org/10.1080/17483107.2023.2277821
5. Jönsson, K.E., Ornstein, K., Christensen, J., Eriksson, J.: A reminder system for independence in dementia care: A case study in an assisted living facility. In: ACM International Conference Proceeding Series, pp. 176–185. Association for Computing Machinery (2019)
6. Mettouris, C., et al.: eSticky: An advanced remote reminder system for people with early Dementia. SN Comput Sci. **4** (2023). https://doi.org/10.1007/s42979-023-01768-3
7. McGoldrick, C., Crawford, S., Evans, J.J.: MindMate: a single case experimental design study of a reminder system for people with dementia. Neuropsychol. Rehabil. **31**, 18–38 (2021). https://doi.org/10.1080/09602011.2019.1653936

8. Sanchez, A.A., Lai, J., Ye, B., Mihailidis, A.: Enhancing communication and autonomy in dementia through technology: navigating home challenges and memory aid usage. Gerontechnology. **23**, 1–11 (2024). https://doi.org/10.4017/gt.2024.23.1.880.06
9. Tay, N.C., Connie, T., Ong, T.S., Teoh, A.B.J., Teh, P.S.: A review of abnormal behavior detection in activities of daily living. IEEE Access. **11**, 5069–5088 (2023). https://doi.org/10.1109/ACCESS.2023.3234974
10. Lai, J., Ye, B., Mihailidis, A.: Anomaly detection technologies for dementia care: monitoring goals, sensor applications, and trade-offs in home-based solutions—a narrative review. J. Appl. Gerontol. (2025). https://doi.org/10.1177/07334648251357031
11. Saumya, S., Singh, J.P.: Spam review detection using LSTM autoencoder: an unsupervised approach. Electron. Commer. Res. **22**, 113–133 (2022). https://doi.org/10.1007/s10660-020-09413-4
12. Lachekhab, F., Benzaoui, M., Tadjer, S.A., Bensmaine, A., Hamma, H.: LSTM-Autoencoder Deep Learning Model for Anomaly Detect. Electric Motor. Energies **17**, (2024). https://doi.org/10.3390/en17102340
13. Wei, Y., Jang-Jaccard, J., Xu, W., Sabrina, F., Camtepe, S., Boulic, M.: LSTM-autoencoder-based anomaly detection for indoor air quality time-series data. IEEE Sens. J. **23**, 3787–3800 (2023). https://doi.org/10.1109/JSEN.2022.3230361
14. Liu, P., Sun, X., Han, Y., He, Z., Zhang, W., Wu, C.: Arrhythmia classification of LSTM autoencoder based on time series anomaly detection. Biomed. Signal Process Control. **71**, (2022). https://doi.org/10.1016/j.bspc.2021.103228
15. Said Elsayed, M., Le-Khac, N.A., Dev, S., Jurcut, A.D.: Network Anomaly Detection Using LSTM Based Autoencoder. In: Q2SWinet 2020 - Proceedings of the 16th ACM Symposium on QoS and Security for Wireless and Mobile Networks, pp. 37–45. Association for Computing Machinery, Inc (2020)
16. Snowball, E., et al.: Engaging people with lived experience of dementia in research meetings and events: insights from multiple perspectives. Front. Dementia. **3**, (2024). https://doi.org/10.3389/frdem.2024.1421737
17. Madara Marasinghe, K.: Assistive technologies in reducing caregiver burden among informal caregivers of older adults: a systematic review (2016)
18. Sriram, V., Jenkinson, C., Peters, M.: Carers' experience of using assistive technology for dementia care at home: A qualitative study. BMJ Open. **10**, (2020). https://doi.org/10.1136/bmjopen-2019-034460
19. McMurray, J., et al.: The importance of trust in the adoption and use of intelligent assistive technology by older adults to support aging in place: Scoping review protocol. JMIR Res Protoc. **6**, (2017). https://doi.org/10.2196/resprot.8772
20. Persson, M., Redmalm, D., Iversen, C.: Caregivers' use of robots and their effect on work environment–a scoping review. J. Technol. Hum. Serv. **40**, 251–277 (2022). https://doi.org/10.1080/15228835.2021.2000554
21. Belmonte-Hernández, A., Semerci, Y.C., Proença, J.P., Romera-Giner, S.: PROCare4Life: An integrated care platform to improve the quality of life of Parkinson's and Alzheimer's patients. In: ACM International Conference Proceeding Series. pp. 358–364. Association for Computing Machinery (2022)
22. Rose, K.M., Lorenz, R.: Sleep disturbances in dementia: what they are and what to do. J. Gerontol. Nurs. **36**, 9–14 (2010). https://doi.org/10.3928/00989134-20100330-05
23. Moldovan, D., et al.: Big Data Analytics for the Daily Living Activities of the People with Dementia. In: 2018 IEEE 14th International Conference on Intelligent Computer Communication and Processing (ICCP), pp. 175–181 (2018)
24. Kim, K., Lee, S., Kim, S., Kim, J., Shin, D.: Sensor-based deviant behavior detection system using deep learning to help dementia caregivers. IEEE Access. **8**, 136004–136013 (2020). https://doi.org/10.1109/ACCESS.2020.3011654

25. Morresi, N., et al.: Measuring Behaviour of People With Dementia Using a Non-Invasive Sensor Network. In: 2024 IEEE International Workshop on Metrology for Industry 4.0 & IoT (MetroInd4.0 & IoT), pp. 339–343 (2024)

26. Chikhaoui, B., Ye, B., Mihailidis, A.: Ensemble Learning-Based Algorithms for Aggressive and Agitated Behavior Recognition. In: García, C.R., Caballero-Gil, P., Burmester, M., Quesada-Arencibia, A. (eds.) Ubiquitous Computing and Ambient Intelligence, pp. 9–20. Springer International Publishing, Cham (2016)

27. Köhler, S., Perry, J., Biernetzky, O.A., Kirste, T., Teipel, S.J.: Ethics, design, and implementation criteria of digital assistive technologies for people with dementia from a multiple stakeholder perspective: a qualitative study. BMC Med. Ethics. **25** (2024). https://doi.org/10.1186/s12910-024-01080-6

28. Mulvenna, M., et al.: Views of caregivers on the ethics of assistive technology used for home surveillance of people living with Dementia. Neuroethics **10**, 255–266 (2017). https://doi.org/10.1007/s12152-017-9305-z

29. Alotaibi, F., et al.: Internet of Things-driven Human Activity Recognition of Elderly and Disabled People Using Arithmetic Optimization Algorithm with LSTM Autoencoder. J. Disab. Res. **2**, (2023). https://doi.org/10.57197/jdr-2023-0038

30. Hou, B., Yang, J., Wang, P., Yan, R.: LSTM-based auto-encoder model for ecg arrhythmias classification. IEEE Trans. Instrum. Meas. **69**, 1232–1240 (2020). https://doi.org/10.1109/TIM.2019.2910342

31. Jung, G., Choi, S.Y.: Forecasting foreign exchange volatility using deep learning autoencoder-LSTM techniques. Complexity. 2021, (2021). https://doi.org/10.1155/2021/6647534

32. Arifoglu, D., Bouchachia, A.: Abnormal behaviour detection for dementia sufferers via transfer learning and recursive auto-encoders. In: 2019 IEEE International Conference on Pervasive Computing and Communications Workshops (PerCom Workshops), pp. 529–534 (2019)

33. Fortin-Simard, D., Bilodeau, J.S., Gaboury, S., Bouchard, B., Bouzouane, A.: Method of recognition and assistance combining passive RFID and electrical load analysis that handles cognitive errors. Int. J. Distrib. Sens. Netw. (2015). https://doi.org/10.1155/2015/643273

34. Nam Ha, K., Chang Lee, K., Lee, S.: Development of PIR sensor based indoor location detection system for smart home. In: SICE-ICASE International Joint Conference, p. 2162. IEEE (2006)

35. Mao, H.F., Chang, L.H., Yao, G., Chen, W.Y., Huang, W.N.W.: Indicators of perceived useful dementia care assistive technology: caregivers' perspectives. Geriatr. Gerontol. Int.Gerontol. Int. **15**, 1049–1057 (2015). https://doi.org/10.1111/ggi.12398

36. Cook, D.J., Crandall, A.S., Thomas, B.L., Krishnan, N.C.: CASAS: a smart home in a box. Computer (Long Beach Calif). **46**, 62–69 (2013). https://doi.org/10.1109/MC.2012.328

BabaSpeech: A Deep Learning-Based Translation of Sign Language Into Lingala Text and Speech for Deaf-Mute Inclusivity

Maurice T. Mukungu[1,4], Alidor M. Mbayandjambe[1,2], Tasho Tashev[3], Kyandoghere Kyamakya[5], and Selain K. Kasereka[1,3,4(✉)]

[1] Department of Mathematics, Statistics and Computer Science, University of Kinshasa, 190 Kinshasa XI, Kinshasa, Democratic Republic of Congo
{alidor.mbayandjambe,selain.kasereka}@unikin.ac.cd
[2] IFI-International School, Vietnam National University, Hanoi, Vietnam
[3] Faculty of English Engineering Studies, Technical University of Sofia, 1000 Sofia, Bulgaria
t_tashev@tu-sofia.bg
[4] ABIL Research Center, 190 Kinshasa XI, Kinshasa, Democratic Republic of Congo
m.mukungu@abil.ac.cd
[5] Institute of Smart Systems Technologies, University of Klagenfurt, Klagenfurt am Worthersee 9020, Klagenfurt, Austria
kyandoghere.kyamakya@aau.at

Abstract. Communication is essential for social and educational inclusion, but it remains a challenge for deaf and mute individuals, especially in the Democratic Republic of the Congo (DRC) and neighboring countries where Lingala is widely spoken. Despite advances in artificial intelligence, existing sign language translation systems have focused mainly on languages such as English and French, failing to address the needs of African language speakers. According to the World Federation of the Deaf (WFD), approximately 70 million deaf individuals live worldwide, with 80% residing in developing countries. These individuals use more than 300 distinct sign languages, each with linguistic structures independent of spoken languages. This paper proposes an intelligent sign language translation system for Lingala, leveraging deep learning techniques. The model is built on a fine-tuned EfficientNet-B3 architecture combined with Long Short-Term Memory (LSTM) for temporal gesture modeling, enabling real-time translation of hand movements into text and speech. This approach enhances accessibility for deaf and mute individuals proficient in Lingala but with limited access to formal education. The results show that the system achieves over 91.8% accuracy, demonstrating its effectiveness for integration into educational programs, language training centers, and media platforms. This work contributes to reducing communication barriers and promoting linguistic accessibility for deaf-mute communities in the DRC and neighboring countries.

Keywords: Sign Language · Communication · Accessibility · Inclusive Education · Deaf-Mute Communities · Artificial Intelligence · Lingala

© The Author(s), under exclusive license to Springer Nature Singapore Pte Ltd. 2025
S. S. Khan et al. (Eds.): IJCAI 2025, CCIS 2620, pp. 91–105, 2025.
https://doi.org/10.1007/978-981-95-0568-5_7

1   Introduction

Sign language serves as a vital bridge for communication among deaf individuals, yet accessibility and awareness remain significant challenges worldwide. According to the WHO, more than 1.5 billion people globally live with hearing impairments, many of whom lack access to inclusive technologies [1]. In the DRC, deaf and mute individuals continue to face persistent communication barriers in education, public services, and media [2]. Despite the critical role of sign language, limited societal awareness continues to hinder the inclusion of the deaf community [3,4]. Recent studies have shown that children with hearing impairments are particularly vulnerable, as early access to communication tools is crucial for their development [5–7]. Without inclusive solutions, they are at risk of both educational and social exclusion. Advances in Artificial Intelligence (AI), particularly CNN-based computer vision, have opened new opportunities for real-time gesture recognition [8].

This study proposes a real-time sign language translation system capable of converting dynamic gestures into Lingala text and speech. By integrating EfficientNet-B3 for spatial feature extraction [9] and an LSTM network for temporal modeling [10], the system ensures accurate, efficient, and culturally adaptive translation. Targeted for use in schools, mobile applications, and public media, this technology seeks to foster inclusive communication, promote civic engagement, and enhance accessibility for deaf-mute individuals, ultimately reinforcing their participation in society. The study contributes to the advancement of AI-driven accessibility solutions by demonstrating the feasibility of a real-time sign language translation system tailored for resource-constrained environments.

Effective communication is a fundamental right, yet deaf-mute individuals in DRC face significant barriers due to the lack of accessible sign language translation technologies tailored to local linguistic and cultural contexts. Recent efforts in the DRC aim to institutionalize Congolese Sign Language (LSC), with a 6,000-word dictionary currently under governmental review, in line with legal rights outlined in Organic Law No. 22/003 [11]. While this marks an important step toward linguistic inclusion, existing sign language technologies predominantly focus on English or French, leaving a critical gap for Lingala, a widely spoken language in the region. Furthermore, many available systems rely on expensive hardware or are limited to static gesture recognition, restricting their usability in low-resource environments.

To guide this investigation, the following research questions have been formulated.

– *How can real-time AI translate sign language gestures into Lingala to support inclusive communication in the DRC?*
– *Which deep learning methods enable accurate, lightweight, and culturally adaptive recognition in low-resource income settings?*

Main Contributions:

- Introduces an innovative approach to bridging linguistic accessibility, presenting the first AI system designed to translate sign language into Lingala text and speech.
- Proposes a hybrid approach for dynamic gesture recognition, integrating a Lightweight EfficientNet-B3 backbone with LSTM-based temporal modeling.
- Presents a transformative deployment strategy tailored for educational and public sectors in resource-constrained environments, with a focus on inclusion for deaf-mute individuals.

The remainder of this paper is organized as follows. Section 2 presents sign language recognition, communication challenges, and reviews related studies. Section 3 outlines the methodology adopted to carry out this study. Section 4 details the experiments and results. Finally, Sect. 5 presents the conclusion and future works.

2 Related Works

2.1 Sign Language Recognition and Communication Challenges

Sign language is essential for deaf communication, yet limited understanding among hearing individuals creates inclusion challenges. AI-driven solutions enable real-time translation of sign language into text and speech, bridging this gap. Sign language functions as a complete linguistic system with unique grammar, syntax, and vocabulary. Deafness varies in severity and may result from genetic factors, infections, or prolonged noise exposure [1,12]. Despite advancements, automated sign language recognition faces challenges such as gesture variability, dialect differences, and structural complexity. High-quality, annotated datasets are crucial for enhancing machine learning models' accuracy [4,13,13]. AI improves accessibility through real-time translation, automatic subtitles, and enhanced sign language education. In Central Africa, Lingala and LSC lack assistive technologies due to standardization and resource limitations. Government efforts to formalize LSC and create a dictionary reflect progress toward linguistic inclusion [2,11]. In recent years, numerous studies have been conducted to bridge communication barriers faced by deaf-mute individuals. These efforts have led to the development of various sign language recognition systems employing advanced computer vision and artificial intelligence techniques.

2.2 Vision-Based Sign Language Recognition Systems

Recent vision-based systems for sign language recognition use cameras and deep learning models like Convolutional Neural Networks (CNN) to translate gestures into text or speech. Tools such as SignAll and SignAloud demonstrate real-time translation using visual and sensor data, though challenges remain in accuracy, latency, and environmental sensitivity [14,15]. Advances like MediaPipe combined with CNNs have achieved up to 99.95% accuracy in ASL alphabet recognition, highlighting the potential of lightweight, high-performance solutions [10].

2.3 Sensor-Based Sign Language Recognition Systems

Sensor-based systems involve wearable devices, such as gloves equipped with sensors, to capture hand and finger movements. These systems offer high accuracy and real-time processing capabilities. However, they often come with higher costs and may be less comfortable for users. The SignAloud system is a notable example, utilizing gloves to detect hand gestures and translating them into speech [15].

2.4 Multi-Modal and AI-Powered Approaches

Multi-modal approaches enhance sign language recognition by integrating visual and linguistic data. Recent systems, such as those by Li et al. and Avina et al., combine computer vision with NLP or deep learning to improve translation accuracy and real-world usability [16,17]. In particular, Sign2GPT utilizes large-scale pre-trained vision language models with lightweight adapters to perform gloss-free translation, effectively addressing data scarcity and computational limitations.

Table 1 compares the key modules of gesture recognition: EfficientNetB3 for spatial feature extraction, LSTM for temporal modeling, and MediaPipe for real-time hand landmark detection.

3 Methodology

In this section, we detail the design of our intelligent sign language translation system, which integrates deep learning and computer vision techniques. The methodology encompasses data acquisition and preprocessing, model architecture selection and fine-tuning, as well as training and evaluation procedures.

3.1 Data Collection and Preprocessing

We constructed a dataset of sign language video sequences annotated with semantic gesture labels, captured in real time using a webcam (640×480 resolution). Using MediaPipe, hand landmarks were detected and cropped to form regions of interest. Videos were decomposed into frames, resized to 224×224 pixels, and normalized to the $[0, 1]$ range.

Data augmentation techniques such as flips, rotations, zoom, and brightness adjustments were applied. Finally, frames were grouped into fixed-length sequences to model temporal dependencies. This preprocessing ensured high-quality and diverse inputs for robust model training.

3.2 System Architecture

The overall architecture of *BabaSpeech* is designed to enable real-time translation of LSC gestures into Lingala text and speech. The system is composed of the following sequential modules:

Table 1. Technical Comparison of the Main Components Used in the Gesture Recognition Pipeline

Criterion	EfficientNetB3 [9]	LSTM [18]	MediaPipe [19]
Role in the Study	Spatial visual feature extraction from gesture images	Temporal modeling of vector sequences to capture gesture dynamics	Detection of hand keypoints for skeletal representation
Nature of Processed Data	Static RGB images	Temporal sequences of vectors (from CNN or MediaPipe)	RGB images $\rightarrow$ coordinates (x, y, z) of 21 hand landmarks
Mathematical Foundation	2D convolutions, MBConv blocks, Swish activation, compound scaling (ϕ)	Recurrent operations + gating mechanisms (input, forget, output), memory cell (C_t)	CNN-based regression predicting 3D landmarks from RGB inputs
Key Equations	– MBConv: depthwise conv + squeeze-and-excitation – Swish: $Swish(x) = x \cdot \sigma(x)$	See LSTM formulas: f_t, i_t, o_t, C_t, h_t (gates and state update)	N/A (optimized black-box model for real-time inference)
Advantages	– Excellent accuracy/efficiency tradeoff – Scalable with compound scaling – Pre-trained on ImageNet	– Captures long-term dependencies – Mitigates vanishing gradient issues – Robust for sequential modeling	– Real-time performance – Lightweight and easy to integrate – No need for supervised training
Limitations	– Less suited for sequential data – Requires a trained backbone	– Slower training – Susceptible to overfitting on short sequences	– Sensitive to lighting conditions – Not optimized for multitask use
Output Type	Feature vectors (per image)	Aggregated temporal vectors for classification	List of 21 hand keypoints (x, y, z) per frame
Use in Our Study	Encodes individual gesture images into feature vectors	Classifies gesture sequences based on extracted vectors	Alternative input to image-based models, enabling skeletal modeling

1. **Hand Landmark Extraction:** A webcam captures continuous video input at a resolution of 640×480 pixels. Each frame is processed by the MediaPipe framework, which detects 21 hand keypoints per frame. These keypoints represent the (x, y, z) coordinates of specific hand joints.

2. **Frame Preprocessing:** The regions of interest (ROI) around the detected hands are extracted and resized to 224×224 pixels. Pixel values are normalized to the $[0, 1]$ range as shown in Eq. 1:

$$I_{\mathrm{norm}} = \frac{I}{255} \tag{1}$$

3. **Feature Extraction:** Each preprocessed frame is passed through an EfficientNet-B3 model, which generates a high-dimensional spatial feature vector. The model leverages compound scaling to balance depth, width, and resolution as demonstrated in Eq. 2 and Eq. 3:

$$d = \alpha^{\phi}, \quad w = \beta^{\phi}, \quad r = \gamma^{\phi} \tag{2}$$

subject to the constraint:

$$\alpha \cdot \beta^2 \cdot \gamma^2 \approx 2 \tag{3}$$

where ϕ is the scaling coefficient and α, β, γ are constants defined during model design.

4. **Temporal Modeling:** Sequences of 16 consecutive frames are grouped and passed into a LSTM network, which captures the temporal dependencies inherent in dynamic sign gestures. The LSTM processes a sequence of feature vectors $\{\mathbf{x}_1, \mathbf{x}_2, \ldots, \mathbf{x}_{16}\}$ to generate a temporally-aware representation $\mathbf{h}_t$.

5. **Classification:** The output from the final LSTM cell is passed through a fully connected dense layer to classify the gesture into one of $C = 9$ predefined classes.

6. **Output Generation:** The predicted gesture label is mapped to its corresponding Lingala translation. This text is then synthesized into audible speech using a Text-To-Speech (TTS) engine to provide real-time auditory feedback.

This modular pipeline ensures low-latency processing (approximately 100ms per inference) and is optimized for deployment in real-time, resource-constrained environments.

3.3   Training and Validation

The model was trained using the cross-entropy loss function, as shwon in Eq. 4 :

$$\mathcal{L}_{\mathrm{CE}} = - \sum_{i=1}^{N} y_i \log(\hat{y}_i) \tag{4}$$

where y_i is the ground truth label and $\hat{y}_i$ is the predicted probability for class i.

We employed 5-fold cross-validation to ensure robust generalization. The dataset was divided into five equally sized subsets, rotating one subset for validation while training on the remaining four. Performance metrics were averaged across all folds.

During training:

– Early stopping was employed based on validation loss to prevent overfitting.
– Batch normalization and dropout were integrated to stabilize and regularize the training process.
– Model checkpoints were saved at the epoch yielding the highest validation accuracy.

3.4   Real-Time Translation Module

Upon successful training, the model was integrated into a real-time translation pipeline:

– **Gesture Capture:** A webcam continuously streams video input, from which frames are extracted.

- **Prediction:** The preprocessed frames are fed sequentially into the EfficientNet-LSTM pipeline to predict the corresponding gesture.
- **Output Generation:** The recognized gesture is translated into text and simultaneously converted into speech using a text-to-speech (TTS) engine, providing auditory feedback.

The reported latency of approximately 100 milliseconds refers specifically to the model inference time measured on a desktop equipped with an NVIDIA RTX 3050 GPU. This includes only the time required to pass the input through the EfficientNet-B3 and LSTM pipeline. The full end-to-end latency, encompassing hand keypoint extraction, frame preprocessing, model inference, and text-to-speech (TTS) synthesis, ranges from 180 milliseconds to 250 milliseconds depending on system workload and hardware performance.

3.5   Model Evaluation and Metrics

To assess the performance of the BabaSpeech model, we adopted standard classification metrics including accuracy, precision, recall, and F1-score. Accuracy measures the overall proportion of correctly classified gestures. Precision indicates how many of the predicted positive classes were actually correct, while recall reflects the ability of the model to capture all true positive instances. The F1-score offers a harmonic mean between precision and recall, providing a balanced measure particularly useful in the presence of class imbalance.

These metrics, computed across all cross-validation folds, provide a comprehensive view of the system's effectiveness in recognizing real-world sign language gestures.

3.6   Overall Approach

The proposed BabaSpeech system follows a modular pipeline that transforms raw video input into Lingala text and speech through a series of real-time processing steps. As shown in Fig. 1, the architecture integrates data collection, preprocessing, gesture recognition based on deep learning,ng, and multimodal output generation.

The system uses MediaPipe to extract 21 hand landmarks from webcam video, creating standardized and augmented input frames. An EfficientNet-B3 and LSTM-based model then recognizes gestures over time, maps them to Lingala phrases, and converts the output into speech using a TTS engine. The end-to-end pipeline runs in real time on standard hardware.

4   Experiments and Results

This section presents the experimental setup, evaluation metrics, and results obtained from training and testing the proposed sign language translation system. A comparative analysis with existing approaches is also provided to highlight the performance of our method.

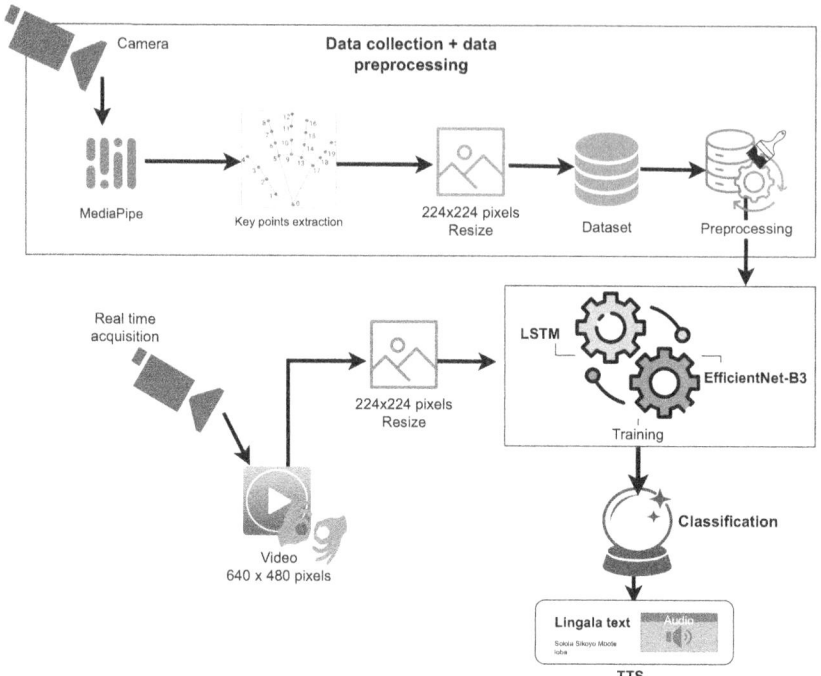

Fig. 1. Overview of the Proposed System

4.1 Dataset Description

The 9 selected gestures represent high-frequency expressions used in everyday communication among Lingala speakers. They were chosen based on their ease of annotation, clarity in execution, and cultural significance. All are static gestures to ensure simplicity and consistent interpretation across participants.

To carry out this study, we created a custom LSC dataset, inspired by national efforts in the DRC to formalize LSC, including a review of 6,000 words sign dictionary [11]. The dataset consists of 3,657 PNG images (224 × 224 pixels) across 9 gesture classes: Lakisa (Show or Demonstrate), Matondo Mingi (Thank you very much), Mbote nayo (Hello to you), Nalingi yo mingi (I love you very much), Naza bien (I am fine), Ndenge nini (How are you?), Pona nini (Why?), Ya sika (New), and Yaka (Come). We used 2,925 images for training and 732 for validation. Each image represents a specific sign, capturing both isolated and sequential expressions. Figure 2 presents a sequence sample used for data collection.

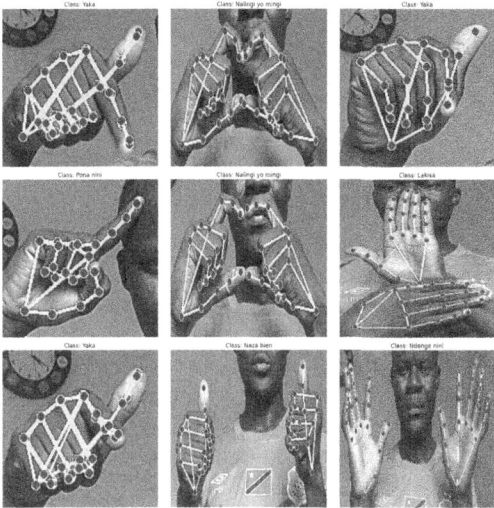

Fig. 2. Examples of LCS Recognition Using Hand Keypoints

4.2   Experimental Setup and Training Configuration

All experiments were conducted on a high-performance workstation (RTX 3050 GPU, 32 GB RAM, Intel i5-12450HX, Lenovo LOQ). Models were implemented in PyTorch following the approach in Sect. 3. Training used the Adam optimizer (initial learning rate 1×10^{-4}), a batch size of 32, and ran for 50 epochs with a cosine annealing scheduler. Cross-entropy loss was used for multiclass classification. Each input consisted of 16 consecutive frames to capture short-term temporal patterns. This configuration ensured efficient training and good generalization across varied gesture sequences.

4.3   Evaluation Metrics

The model performance was assessed using five key metrics. Accuracy (Acc) measures the proportion of correctly predicted gestures relative to the total number of predictions. Precision (P) quantifies the ratio of correctly predicted positive observations to all predicted positives, while Recall (R) evaluates the model's ability to identify all actual positive instances. The F1-Score (F1), defined as the harmonic mean of precision and recall, offers a balanced assessment, particularly valuable in the presence of class imbalance. Finally, Inference Speed, measured in frames per second (FPS), was used to evaluate the model's capacity to operate in real time.

4.4   Results

Tables 2 and 3 present the performance of our models on the test set, with a focus on evaluating the impact of temporal modeling. An ablation study was conducted to measure the contribution of sequential information using an LSTM module. Two configurations were compared:

– **EfficientNet-B3 (baseline)**: Processes each frame independently without temporal modeling.
– **EfficientNet-B3 + LSTM (proposed)**: Frame-level features are fed into an LSTM to capture temporal dependencies.

The classification report for the baseline model Table 2 shows moderate performance, with an overall accuracy of 78% and significant variability across classes. In comparison, the proposed EfficientNet-B3 + LSTM model Table 3 achieves a substantial improvement, reaching an overall accuracy of 91.8% and exhibiting consistently high precision, recall, and F1-scores across almost all classes.

Table 2. Classification Report for EfficientNetB3 (Standard)

Class	Precision	Recall	F1 Score	Support
Lakisa	0.50	1.00	0.67	95
Matondo Mingi	0.78	0.99	0.87	82
Mbote nayo	1.00	1.00	1.00	80
Nalingi yo mingi	1.00	0.39	0.57	76
Naza bien	1.00	0.76	0.87	80
Ndenge nini	0.96	1.00	0.98	79
Pona nini	0.66	0.99	0.79	80
Ya sika	0.00	0.00	0.00	80
Yaka	1.00	0.82	0.90	80
Accuracy			**0.78**	**732**

Table 3. Classification Report for EfficientNetB3 + LSTM

Class	Precision	Recall	F1-Score	Support
Lakisa	0.98	0.66	0.79	95
Matondo Mingi	0.95	0.93	0.94	82
Mbote nayo	1.00	1.00	1.00	80
Nalingi yo mingi	1.00	0.92	0.96	76
Naza bien	1.00	0.90	0.95	80
Ndenge nini	0.66	1.00	0.80	79
Pona nini	0.95	0.99	0.97	80
Ya sika	0.88	0.97	0.92	80
Yaka	1.00	0.94	0.97	80
Accuracy			**0.918**	**732**

These findings underscore the importance of temporal modeling in sign language recognition. By leveraging the sequential nature of gestures, the LSTM layer enables the model to better disambiguate visually similar signs, resulting in more accurate and reliable classification. This demonstrates the effectiveness of integrating temporal context for real-world sign language interpretation systems.

4.5   Comparative Analysis

A comparative analysis was performed against the baseline models commonly used for gesture recognition, namely ResNet50 and MobileNetV2. All models were fine-tuned under the same experimental conditions (Table 4).

Table 4. Results Summary of EfficientNetB3 vs. EfficientNetB3 + LSTM

Metric	EfficientNetB3	EfficientNetB3 + LSTM	Difference
Validation Precision	0.7801	0.9180	+0.1379
Validation Loss	1.8185	0.63685	−1.18165
Training Time (s)	8581.68	3595.53	−4986.15
Relative Improvement	N/A	+17.69%	N/A

The addition of LSTM layers resulted in a 17.69% increase in validation accuracy, reaching 0.9180 compared to 0.7801 for the baseline model.
Recommendation for the Lingala Sign Language Translation System: Use the EfficientNetB3 + LSTM model for improved accuracy, with a reasonable trade-off in training time.

The proposed EfficientNet-B3 + LSTM model outperformed the baselines in all evaluation metrics while maintaining real-time inference capabilities. This highlights its suitability for deployment in real-world sign language translation applications.

4.6   Evaluation Results and Temporal Modeling Impact

The following visualizations illustrate the classification performance of two architectures applied to Lingala sign language recognition, as shown in Fig. 3. By comparing the standard EfficientNetB3 with its LSTM-augmented variant, we assess the impact of temporal modeling on sequential sign interpretation accuracy.

4.7   Validation Accuracy and Training Time Comparison

To evaluate the performance trade-offs between the standard EfficientNetB3 and its LSTM-augmented counterpart, we compare both validation accuracy and training time, as shown in Fig. 4. The results clearly demonstrate that the temporal modeling capability of the LSTM improves classification performance while reducing training time.

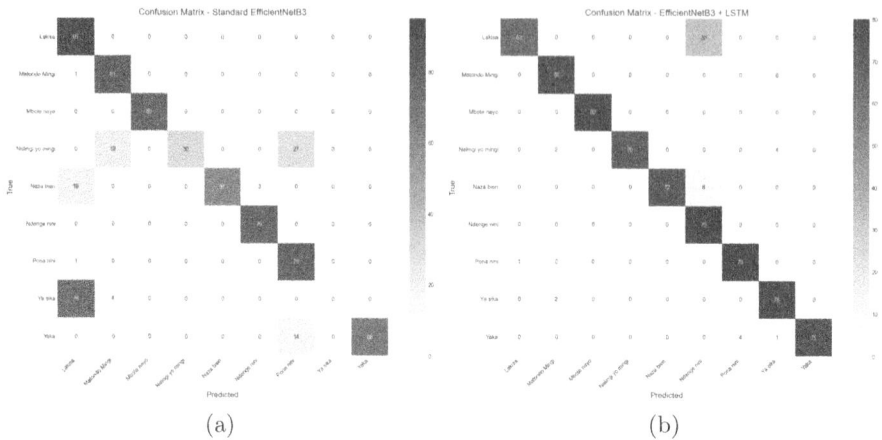

(a)   (b)

Fig. 3. (a) EfficientNetB3 shows strong results on isolated gestures but struggles with sequential ones. (b) Adding LSTM improves recognition of context-dependent gestures.

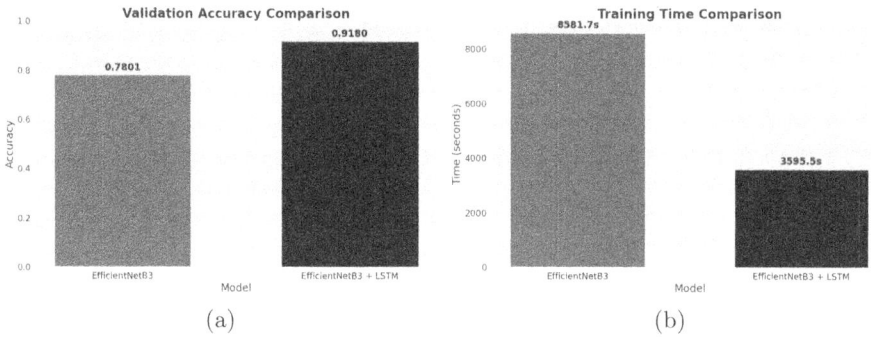

(a)   (b)

Fig. 4. (a) Validation accuracy comparison. EfficientNetB3 + LSTM achieves higher accuracy (0.9180) compared to the standard model (0.7801). (b) Training time comparison. The LSTM-augmented model trains faster (3595.5 s) than the baseline (8581.6 s).

4.8   Gesture Visualization with Keypoint Detection

In this study, we used a hand keypoint detection model to automatically extract joint positions while performing sign language gestures. The visualization below illustrates two examples: the gestures for *Matondo Mingi* and *Naza bien*. Figure 5 shows sample images from different sign categories.

4.9   Discussion

Our study demonstrates the effectiveness of integrating EfficientNetB3 with Long Short-Term Memory (LSTM) networks for the recognition of Lingala Sign Language gestures. As reported in Table 3, the hybrid model achieved a validation accuracy of 91.8%, substantially higher than the 78% obtained by the standalone

Fig. 5. Examples of sign language gestures extracted from a video sequence. The red dots represent hand keypoints detected by the model, enabling accurate motion interpretation. (Color figure online)

EfficientNetB3 model (Table 2). These results highlight the importance of modeling temporal dependencies in sign language, which are effectively captured by the LSTM component.

The confusion matrices in Fig. 3 illustrate significant performance gains for temporal gestures such as "Ya sika" and "Nalingi yo mingi," while a slight degradation was observed in the classification of "Lakisa." Other classes, such as "Mbote nayo," "Ndenge nini," and "Pona nini," remained stable across both configurations. These findings are consistent with previous studies. Huang and Chouvatut [20] achieved 86.25% accuracy using a ResNet-LSTM for Argentine Sign Language, while Kautsar et al. [21] reported 94.67% accuracy for BISINDO recognition with an LSTM-based model. Such results reinforce the effectiveness of combining convolutional feature extractors with temporal sequence models.

Training time analysis in Fig. 4 also indicates a 58.1% reduction in duration when using the hybrid model (3595.5 s) compared to EfficientNetB3 alone (8581.6 s). This efficiency gain stems from freezing pretrained EfficientNetB3 layers, lower-dimensional sequential input to the LSTM, and faster convergence during training.

4.10 Limitations of the Model

While the EfficientNetB3 + LSTM architecture yields promising results, several limitations remain. The dataset lacks diversity in signers and environments, which restricts its generalization. The system currently handles only isolated gestures, missing the complexity of continuous sign language and non-manual cues. Additionally, it is not optimized for low-resource deployment. Furthermore, the system has only been tested on adult Black subjects and has not been evaluated on children or individuals of other racial backgrounds. Additionally, the translation is not bidirectional.

5 Conclusion and Future Works

The institutionalization of LSC marks a significant milestone in promoting linguistic inclusion in the DRC. However, existing sign language technologies primarily cater to English and French, leaving a gap in accessibility for Lingala speakers. Recognizing this challenge, the proposed real-time sign language translation system offers a transformative approach by integrating EfficientNet-B3 for spatial feature extraction and an LSTM network for temporal modeling. Designed for deployment in schools, mobile applications, and public media, this system fosters inclusive communication and enhances civic engagement for deaf-mute individuals.

This study contributes to AI-driven accessibility solutions, demonstrating the feasibility of real-time gesture translation in resource-constrained environments. In addition, the insights gained from optimizing deep learning models for dynamic gesture recognition pave the way for broader applications in multilingual sign language translation systems. As the DRC moves forward with institutionalizing LSC, implementing assistive technologies such as this will be crucial to strengthening equal communication rights, ensuring that linguistic inclusion is both sustainable and impactful.

In the future, we aim to expand this work by developing a system dedicated to learning sign language and integrating its translation into a mobile or desktop application. Additionally, we plan to create a system capable of translating Lingala text and speech into sign language, ensuring full inclusivity for the deaf-mute community. To enhance this effort, future developments should prioritize expanding the dataset, incorporating facial and emotional features, improving inference efficiency through lightweight models, and exploring multimodal integration for more comprehensive sign language interpretation.

Acknowledgments. This work has been realized with financial support by the European Regional Development Fund within the Operational Programme "Bulgarian national recovery and resilience plan", Procedure for direct provision of grants "Establishing of a network of research higher education institutions in Bulgaria", under the Project BG-RRP-2.004-0005 "Improving the research capacity and quality to achieve international recognition and resilience of TU-Sofia".

Code Availability Statement. The code supporting the findings of this study is available at the following repository: https://github.com/FrereAlidor/LSC-Lingala-sign-Efficient-LSTM.

References

1. Deafness and hearing loss - world health organization (who). 26 février 2025
2. Mbayandjambe, A.M., Kasereka, S.K., Kyamakya, K., Ho, V.T.: Enhancing printed lingala script recognition using deep learning techniques. Proc. Comput. Sci. **257**, 111–118 (2025). https://doi.org/10.1016/j.procs.2025.03.017
3. Pribanić, L.: Sign language and deaf education: a new tradition. Sign Lang. Linguist. **9**(1–2), 233–254 (2006)
4. Van Staden, A., Badenhorst, G., Ridge, E.: The benefits of sign language for deaf learners with language challenges. Per Linguam **25**(1) (2011). https://doi.org/10. 5785/25-1-28

5. Beaujard, L., Perini, M.: The role and place of sign language in deaf youth's access to literacy: Contributions of a cross-review of ASL-English and lsf-french research. Front. Commun. **7**, 810724 (2022)
6. Abdulaziz Abdullah Alothman: Language and literacy of deaf children. Psychol. Educ. **58**(1), 799–819 (2021)
7. Herman, R., Rowley, K.: Assessments of sign language development: New insights. *Hrvatska revija za rehabilitacijska istraživanja.* **58**(Special Issue), 98–108 (2022)
8. van Berkel, J.A.N., Kostakos, V.: Artificial intelligence for sign language translation a design science research study. In: IEEE/CVF Conference on Computer Vision and Pattern Recognition (CVPR), 200, pp. 10023–10033, May 24 (2020)
9. Tan, M., Le, Q.V.: Efficientnet: rethinking model scaling for convolutional neural networks. International Conference on Machine Learning (ICML), pp. 6105–6114. PMLR(2019)
10. Kumar, R., Bajpai, Sinha, A.: Mediapipe and CNNs for real-time ASL gesture recognition. arXiv preprint arXiv:2305.05296 (2023)
11. Congolese Press Agency. A dictionary in Congolese sign language is under validation ministry of persons with disabilities (2023). Accessed 2 May 2025
12. What is American sign language (ASL)? 29 octobre 2021
13. Grif, M.G., Kugaevskikh, A.V.: Recognition of deaf gestures based on a bio-inspired neural network. J. Phys.: Conf. Series **1661**(1), 012038 (2020). https://doi.org/10.1088/1742-6596/1661/1/012038
14. Hou, J., et al.: Signspeaker: A real-time, high-precision smartwatch-based sign language translator. In: The 25th Annual International Conference on Mobile Computing and Networking, pp. 1–15, (2019)
15. L MIT. Signaloud: Gloves that transliterate sign language into text and speech (2016)
16. Li, J., et al.: Sign language recognition and translation: a multi-modal approach using computer vision and natural language processing. In: Proceedings of the 14th International Conference on Recent Advances in Natural Language Processing, pp. 658–665 (2023)
17. Avina, V.D., Amiruzzaman, Md., Amiruzzaman, S., Ngo, L.B., Dewan, M.A.A.: An AI-Based Framework for translating american sign language to English and vice versa. Information **14**(10), 569 (2023). https://doi.org/10.3390/info14100569
18. Hochreiter, S., rgen Schmidhuber, J.: Long short-term memory. Neural Comput. **9**(8), 1735–1780 (1997)
19. Lugaresi, C., et al. Mediapipe: A framework for building perception pipelines. arXiv preprint arXiv:1906.08172 (2019)
20. Huang, J., Chouvutut, V.: Video-based sign language recognition via RESnet and LSTM network. J. Imaging **10**(6), 149 (2024)
21. Kautsar, M.D., Hariono, A.M., Akmal, R.: Improving word-level bisindo recognition, Attention vs LSTM (2025)

Scene Invariant Cross Camera Anomaly Detection of Behaviours of Risk in People with Dementia

Pratik K. Mishra[1,2]([✉]) [iD], Andrea Iaboni[2,3] [iD], Bing Ye[2] [iD],
Kristine Newman[4] [iD], Alex Mihailidis[1,2] [iD], and Shehroz S. Khan[1,2] [iD]

[1] Institute of Biomedical Engineering, University of Toronto, Toronto, ON, Canada
`pratik.mishra@mail.utoronto.ca, alex.mihailidis@utoronto.ca,`
`shehroz.khan@uhn.ca`
[2] KITE, Toronto Rehabilitation Institute, Toronto, ON, Canada
`andrea.iaboni@uhn.ca, bing.ye@utoronto.ca`
[3] Department of Psychiatry, Temerty Faculty of Medicine, University of Toronto,
Toronto, ON, Canada
[4] Daphne Cockwell School of Nursing, Toronto Metropolitan University, Toronto,
ON, Canada
`kristine.newman@torontomu.ca`

Abstract. Behavioural and psychological symptoms of dementia, including agitation and aggression, pose substantial health and safety risks in residential care environments. The widespread use of video surveillance in common areas within these facilities offers an opportunity to develop automated systems for detecting behaviours of risk. Such systems can provide real-time alerts to staff, facilitating timely intervention and helping to prevent the escalation of potentially harmful situations. A major challenge for these systems is their adaptability to new environments without a significant drop in their detection performance. We propose noiseCAE to tackle this challenge by utilizing partial masking of scene background to mitigate scene bias and assist the autoencoder model to learn scene-invariant normal behaviour characteristics. This approach helps the autoencoder to generalize better to unseen camera scenes without any additional training. The data from nine individuals with dementia, recorded using three cameras positioned in different hallways of a dementia care unit, was utilized for this study. The generalization performance of noiseCAE was investigated in a cross-camera setting, where noiseCAE performed better than the existing method for three out of six cases. This motivates further research to develop scene-invariant cross-camera behaviours of risk detection systems for people with dementia in care environments.

Keywords: Dementia · Behaviours of risk · Agitation · Cross-camera · Autoencoder · Video Anomaly Detection

1 Introduction

Agitation is a behaviour of risk exhibited by people living with dementia (PwD), and it can manifest in multiple ways, including physical aggression, motor agitation, and verbal aggression [1]. These behaviours pose health and safety risks

© The Author(s), under exclusive license to Springer Nature Singapore Pte Ltd. 2025
S. S. Khan et al. (Eds.): IJCAI 2025, CCIS 2620, pp. 106–116, 2025.
https://doi.org/10.1007/978-981-95-0568-5_8

not only to the PwD but also to fellow residents, caregivers, and staff in long-term care settings. However, the persistent shortage of trained personnel in many care homes makes it difficult to maintain continuous observation of all residents, thereby limiting the staff's capacity to respond promptly to residents' needs and safety [2].

Many long-term care facilities are equipped with CCTV cameras to support safety and security within the unit. However, these camera feeds are often either monitored manually by staff or not actively observed. Despite this, the footage contains rich spatio-temporal information capturing the behaviours and inter-actions of residents, visitors, and staff - data that can be leveraged to train deep learning methods for detecting behaviours of risk [3,4]. While individuals in dementia units may have a known history of agitation, such events occur infre-quently, resulting in a significant imbalance between instances of agitation and normal daily activities [5,6]. This data imbalance renders traditional supervised classification approaches unsuitable, as they require predefined class descriptions and a sufficient number of labelled examples for each class to perform effectively [7,8]. Anomaly detection offers a more viable alternative, enabling methods to learn characteristics of normal behaviour and subsequently flag deviations as potential anomalies during inference [9].

In our previous work [10], we employed a depth-weighted spatio-temporal convolutional autoencoder (depCAE) to identify behaviours of risk in PwD by modelling them as anomalous events. However, a drop in performance was observed when the trained depCAE method was evaluated on a new camera scene. In this work, we propose noiseCAE, an extension to depCAE, to leverage foreground-background separation of a camera scene as part of our scene bias mitigation strategy to improve the generalization performance of the autoen-coder method for a new scene. The moving people and objects in the scene are referred to as foreground, and the stationary environment is referred to as background. Partial masking is performed by adding noise to the background to assist the autoencoder method in becoming less sensitive to the scene and focus-ing on the normal behaviour characteristics in the foreground. We compare the performance of noiseCAE with our previous method depCAE in a cross-camera experimental setup in Sect. 4. To the best of our knowledge, this is the first work that analyzes behaviours of risk detection in PwD in a cross-camera setting.

2   Related Work

In this section, we review existing research focused on detecting behaviours of risk in PwD using video-based approaches, and methods to improve generalization performance in cross-camera video anomaly detection (VAD) settings.

2.1   Behaviours of Risk Detection

Current approaches for detecting behaviours of risk in PwD utilize a range of sensing modalities, including wearable devices [11], computer vision techniques

[3,12], and multimodal or ambient sensing systems [13]. As this study focuses on video data, the following review concentrates on work that either relies solely on video or integrates it with other sensor modalities.

Qiu et al. [13] proposed a multimodal information fusion framework that combined data from pressure sensors, ultrasound sensors, infrared sensors, video cameras, and acoustic sensors. Their system employed a layered classification structure comprising a hierarchical hidden Markov model and a support vector machine. However, the results were based on simulated data, limiting the applicability of the findings to real-world contexts. Chikhaoui et al. [14] developed an ensemble classifier to detect agitation using input from a Kinect camera and an accelerometer. The study involved ten participants instructed to perform six types of agitated or aggressive behaviours, although it was not reported whether the participants were PwD or healthy individuals. Similarly, Fook et al. [12] proposed a method based on a multi-layer probabilistic classifier, integrating a hidden Markov model and a support vector machine. However, the video data captured only a person in bed, and participant demographics were not disclosed. In all these studies, the systems were conceptual and tested in controlled environments, with no validation on real-world patient data, leaving their practical utility and generalizability uncertain. Khan et al. [5] introduced an unsupervised convolutional autoencoder for detecting agitation in PwD using video footage from a specialized dementia unit, based on data from a single participant and a single camera. Mishra et al. [3] applied privacy-preserving VAD methods, incorporating body pose representations and semantic segmentation masks to detect behaviours of risk while safeguarding participant privacy.

2.2   Cross Camera Video Anomaly Detection

Existing VAD approaches typically operate under the implicit assumption that models trained on a given set of videos can be directly applied to unseen test videos. This assumption holds when the training and testing videos originate from the same scene, such as those captured by the same camera under consistent environmental conditions. However, when a model trained on videos from one scene is evaluated on footage from a different scene, performance degrades significantly due to variations in environmental conditions [15]. In cross-camera VAD, the goal is to produce a VAD model that can generalize well to a new target scene with little to no fine-tuning using target scene frames.

Lu et al. [15] proposed a few-shot scene-adaptive anomaly detection framework, where meta-learning was used to facilitate rapid model adaptation to new environments. Lv et al. [16] proposed a dynamic prototype unit that encoded normal behavioural dynamics as prototypes in real time and integrated meta-learning in their method to develop a few-shot normalcy adaptation framework. Georgescu et al. [17] proposed a background-agnostic anomaly detection framework which operated on detected objects, rather than entire frames and used an adversarial learning strategy to generate a set of scene-agnostic, out-of-domain pseudo-abnormal samples. Unlike the above-discussed methods, which circumvent the background and use other strategies to mitigate the scene bias, we

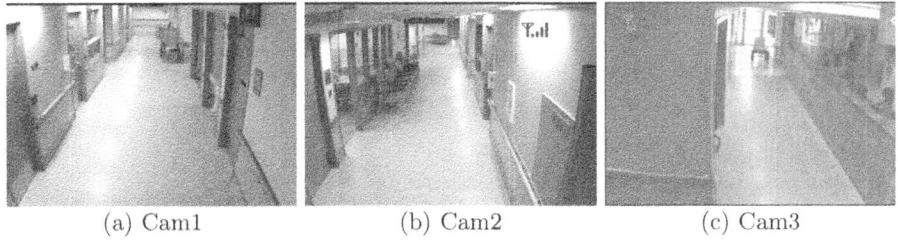

(a) Cam1   (b) Cam2   (c) Cam3

Fig. 1. Camera views from three different hallways in the dementia unit.

follow a direct approach and use partial masking of the background to make the method generalizable to new scenes.

3 Methods

3.1 Dataset Description

The data utilized in this study were collected over two years, from November 2017 to October 2019, at the Specialized Dementia Unit at Toronto Rehabilitation Institute, University Health Network (UHN), Toronto, Canada [5]. The data collection was approved by the research ethics board (UHN REB#14-8483). Video recordings were captured using cameras installed in shared spaces such as hallways, dining areas, and the recreation hall. For the present analysis, we focused on behaviours of risk events involving nine dementia participants captured by three cameras across different hallways of the dementia unit (please refer Fig. 1). Table 1 provides the demographic information of the participants. In addition to the behaviours of risk episodes, the selected video feed also included normal daily activities involving other residents, nursing staff, and visitors within the unit. Informed consent was obtained from the substitute decision-makers of all participating individuals. Additionally, staff members provided written consent to permit video recording within the unit.

3.2 Dataset Preprocessing

The raw video data used in this study had an initial resolution of 352×240 pixels, a frame rate of 30 frames per second (fps), and a bit depth of 24. To reduce computational overhead, frames were sampled at 15 fps by selecting every alternate frame. These frames were then preprocessed by converting them to grayscale, normalizing the pixel values to the [0, 1] range, and resizing each frame to 64×64 pixels. Subsequently, the preprocessed frames were segmented into non-overlapping 5-second windows, each comprising 75 consecutive frames. Independent training and testing sets were curated for each of the three cameras. Both the training and testing sets were subjected to the same preprocessing steps. The training sets contained only examples of normal daily activities, while the test sets included both normal and annotated behaviours of risk windows. The

Table 1. Participants' demographic information.

Number of Participants	9
Median age of participants (years)	82
Mean age of participants (years)	81.22
Standard deviation of the age of participants (years)	8.12
Range of age of participants (years)	66–93
Gender	Males (3) Females (6)

Table 2. Size of training and test sets (in minutes).

	Cam1	Cam2	Cam3
Train set	1225.25	1242.33	1205.66
Test set	209.16	291.92	96
Normal behaviour (Test set)	194.08	270.5	89.83
Behaviours of risk (Test set)	15.08	21.42	6.17

average duration of training data across the three cameras was approximately 1224.41 min. The testing sets comprised a range of behaviours of risk events from nine participants, along with normal behaviours. Details on the size of the training and test sets for each camera are provided in Table 2. These curated testing sets were used to evaluate the performance of the proposed method in Sect. 4.

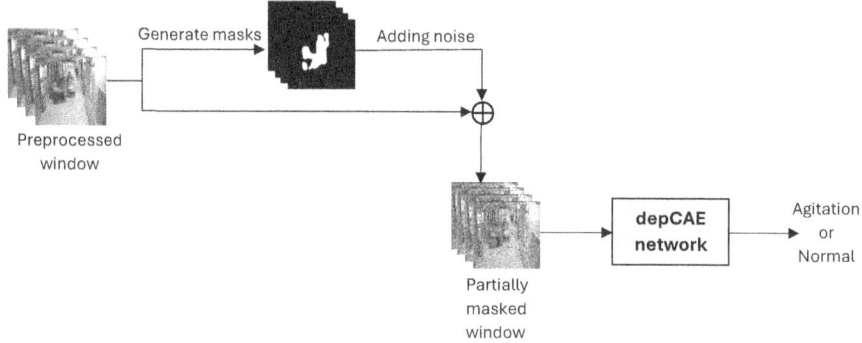

Fig. 2. Flow diagram of noiseCAE for scene-invariant behaviours of risk detection.

3.3 Behaviours of Risk Detection

In the context of anomaly detection, 3D convolutional autoencoders are trained exclusively on data representing normal behaviours. By minimizing reconstruction error during training, these models learn to effectively reconstruct the

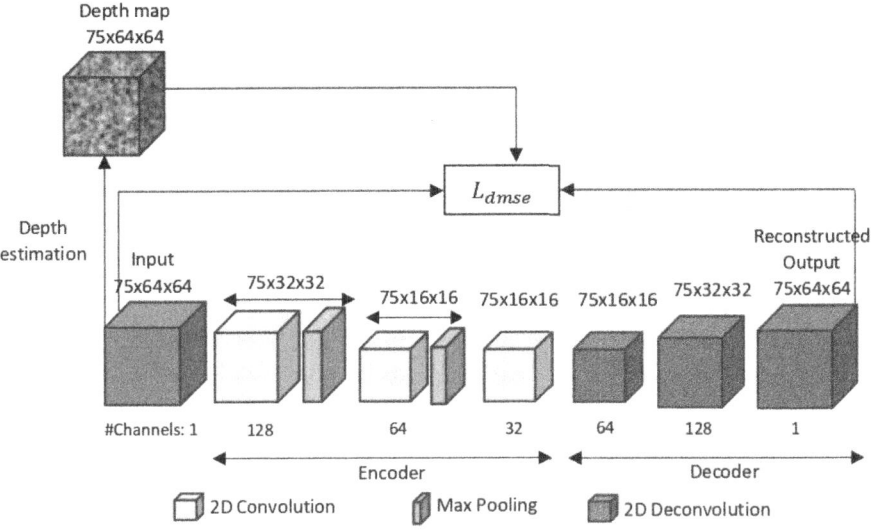

Fig. 3. depCAE architecture to detect behaviours of risk in PwD as anomaly [10].

spatio-temporal patterns characterized by normal behaviour. At test time, anomalous inputs - unseen during training - are expected to yield higher reconstruction errors, enabling their identification as deviations from the norm. In our previous study [10], we employed a depth-weighted 3D convolutional autoencoder, referred as depCAE, to reduce false positive rate during detection of behaviours of risk events in PwD. Further, we investigated the generalization performance of depCAE in a cross-camera setting. A drop in performance of depCAE was observed when tested on a camera different from the training camera due to scene bias resulting from variation in environmental factors such as background, camera angle, and lighting across cameras. In this work, we propose noiseCAE, a method to mitigate this scene bias to improve the cross-camera generalization performance of depCAE. Our proposed approach involves adding noise to the background scene in the input frame windows to achieve partial masking of the background to mitigate the bias resulting from environmental factors. The partial masking is done during training only, where the partially masked frame windows representing normal behaviour are passed to our previous depCAE [10] method to learn scene-invariant normal characteristics. During testing, the preprocessed windows are directly passed as input to the depCAE method without partial masking.

DeepLabv3 [18] and OpenCV background subtraction were used to generate background masks from the previously preprocessed windows. Further, noise was added to the background to generate partially masked windows, which was then fed as input to the depCAE method to detect behaviours of risk as anomalies. The flow diagram of the noiseCAE methodology is presented in Fig. 2. The depCAE [10] method follows an encoder-decoder architecture, where the encoder consisted of 2D convolution and max-pooling layers, with a kernel size of $(1 \times 3 \times 3)$, stride $(1 \times 1 \times 1)$ and padding $(0 \times 1 \times 1)$, followed by batch normalization and ReLU operations. The decoder used 2D deconvolution layers (2D transposed convolution operation) with a kernel size of $(1 \times 3 \times 3)$, strides $(1 \times 1 \times 1)$, $(1 \times 2 \times 2)$, $(1 \times 2 \times 2)$ and paddings $(0 \times 1 \times 1)$, $(0 \times 1 \times 1)$, $(0 \times 1 \times 1)$ for first, second, and third 2D deconvolution layers, respectively. The network architecture of the depCAE is presented in Fig. 3.

Threshold Determination: A key challenge in anomaly detection lies in selecting an appropriate operating threshold, as anomalous samples are absent from the training set. To address this, we adopted the strategy employed in our previous work [19], where we used the windows with high reconstruction error in the training set as proxy outliers to create a proxy validation set containing both normal and proxy outlier instances. The reconstruction errors corresponding to the training windows were treated as potential thresholds. For each candidate threshold, we computed the F1-score on the proxy validation set using cross-validation. The reconstruction error that yielded the highest F1-score was selected as the operating threshold. This threshold was subsequently applied during testing to identify behaviours of risk.

Evaluation Metrics: In the presence of imbalanced data, conventional performance metrics such as accuracy can yield misleadingly optimistic and inflated evaluations of model performance [20]. Hence, we use Matthew's Correlation Coefficient (MCC), F1-score, and area under the receiver operating characteristic (AUROC) curve to evaluate the performance of our model. MCC provides a comprehensive evaluation by incorporating all four confusion matrix components—true positives (TP), true negatives (TN), false positives (FP), and false negatives (FN)—making it particularly suitable for assessing performance in imbalanced classification settings. The F1-score, defined as the harmonic mean of precision and recall, offers a balanced measure that accounts for true positive rate (TPR), false positive rate (FPR), and false negative rate (FNR) simultaneously. Both F1-score and MCC are threshold-dependent metrics, assessing model performance at a specific decision boundary. In contrast, the AUROC provides a threshold-independent evaluation by measuring the model's ability to distinguish between classes across a continuum of thresholds, summarizing the trade-off between TPR and FPR. We additionally report several reference metrics, including TPR, true negative rate (TNR), FPR, FNR, gmean ($\sqrt{TPR * TNR}$), precision, recall, and specificity. These metrics utilize the reconstruction error scores during testing to identify behaviours of risk as anomalies.

4   Results and Discussion

In this section, we evaluate and compare the generalization performance of noise-CAE with depCAE in a cross-camera setting. As part of the experimental setup, the depCAE and noiseCAE methods are tested on one of the three unseen cameras after being individually trained on the remaining two cameras. Table 3

Table 3. Cross-camera analysis when trained on different cameras and tested on Cam1.

	Trained on Cam2		Trained on Cam3	
	depCAE	noiseCAE	depCAE	noiseCAE
TPR	0.862	0.718	0.558	0.575
TNR	0.653	0.704	0.824	0.799
FPR	0.347	0.296	0.176	0.201
FNR	0.138	0.282	0.442	0.425
Gmean	0.75	0.711	0.678	0.677
Precision	0.162	0.159	0.198	0.182
Recall	0.862	0.718	0.558	0.575
Specificity	0.653	0.704	0.824	0.799
MCC	**0.274**	0.233	**0.246**	0.23
F1-score	**0.272**	0.26	**0.292**	0.276
AUROC	**0.796**	0.767	**0.816**	0.783

Table 4. Cross-camera analysis when trained on different cameras and tested on Cam2.

	Trained on Cam1		Trained on Cam3	
	depCAE	noiseCAE	depCAE	noiseCAE
TPR	0.848	0.525	0.638	0.412
TNR	0.585	0.82	0.75	0.888
FPR	0.415	0.18	0.25	0.112
FNR	0.152	0.475	0.362	0.588
Gmean	0.704	0.656	0.692	0.605
Precision	0.139	0.188	0.168	0.225
Recall	0.848	0.525	0.638	0.412
Specificity	0.585	0.82	0.75	0.888
MCC	**0.22**	**0.22**	0.226	**0.229**
F1-score	0.239	**0.277**	0.266	**0.291**
AUROC	0.77	**0.79**	0.774	**0.783**

Table 5. Cross-camera analysis when trained on different cameras and tested on Cam3.

	Trained on Cam1		Trained on Cam2	
	depCAE	noiseCAE	depCAE	noiseCAE
TPR	0.27	0.378	0.27	0.419
TNR	0.922	0.84	0.913	0.856
FPR	0.078	0.16	0.087	0.144
FNR	0.73	0.622	0.73	0.581
Gmean	0.499	0.564	0.497	0.599
Precision	0.192	0.139	0.175	0.167
Recall	0.27	0.378	0.27	0.419
Specificity	0.922	0.84	0.913	0.856
MCC	**0.165**	0.141	0.15	**0.183**
F1-score	**0.225**	0.204	0.21	**0.238**
AUROC	**0.676**	0.619	0.613	**0.623**

presents the testing results of depCAE and noiseCAE methods on Cam1 after being separately trained on Cam2 and Cam3. Similarly, Tables 4 and 5 present the results after the train and test cameras are changed to cover all possible scenarios. The best results for the evaluation metrics MCC, F1-score and AUROC are highlighted in bold. The following observations can be made from the three tables.

1. The noiseCAE method performed better for three out of six cases, while the depCAE method performed better for the remaining three cases.
2. When trained on cameras 1 and 3, and tested on Cam2, which has the most number of behaviour of risk windows, the noiseCAE method performed better than depCAE.
3. When tested on Cam3, which has the least number of behaviour of risk windows, noiseCAE performed better for one case, while depCAE performed better for the other.

The noiseCAE model performed better than the depCAE for the camera with the most number of behaviours of risk windows and similar to depCAE for camera with the least number of behaviours of risk windows. Overall, noiseCAE was able to improve the generalization performance over depCAE for half of the total cases.

4.1   Conclusion and Future Work

The rising number of individuals with dementia, coupled with persistent under-staffing in care facilities, presents significant challenges for ensuring resident safety. Behavioural and psychological symptoms associated with dementia can result in incidents that compromise the well-being of patients, staff, and care-givers. Harnessing existing video surveillance infrastructure offers an opportunity to develop advanced deep learning systems capable of detecting such behaviours in real time, thereby supporting timely interventions, preventing harm, and improving the overall quality of care. However, training deep learning systems for each new environment can be an expensive and time-consuming task. In this direction, we proposed an approach to improve the generalization performance of a 3D autoencoder-based deep learning system using partial masking of the background to learn scene-invariant normal behaviour characteristics. This helped to mitigate scene bias during the training of the autoencoder. The proposed approach performed better than our earlier method for half of the total cases. While the outcome is not optimum, it presents an opportunity to develop and improve scene-invariant cross-camera behaviours of risk detection systems for PwD. The performance of the proposed system also depends upon the quality of the generated foreground masks. Future work can involve using methods to segment motion from videos [21] to generate better quality foreground masks. To enhance the cross-camera generalization, fine-tuning on the target camera using few-shot learning and transfer learning techniques can be employed. These approaches allow the model to adapt to scene-specific variations with minimal additional data, thereby improving its performance in previously unseen environments.

Acknowledgments. This work was supported by Alzheimer's Association, Natural Sciences and Engineering Research Council, UAE Strategic Research Grant, Walter and Maria Schroeder Institute for Brain Innovation and Recovery.

References

1. Cohen-Mansfield, J.: Conceptualization of agitation: results based on the Cohen-mansfield agitation inventory and the agitation behavior mapping instrument. Int. Psychogeriatr. **8**(S3), 309–315 (1997)
2. of Long-Term Care, M.: Long-term care staffing study (July 202). https://www.ontario.ca/page/long-term-care-staffing-study. Accessed 1 Mar 2023
3. Mishra, P.K., Iaboni, A., Ye, B., Newman, K., Mihailidis, A., Khan, S.S.: Privacy-protecting behaviours of risk detection in people with dementia using videos. Biomed. Eng. Online **22**(1), 1–17 (2023)
4. Mishra, P.K., Iaboni, A., Ye, B., Newman, K., Mihailidis, A., Khan, S.: Detection of behaviors of risk exhibited by people with dementia with privacy protecting videos. Alzheimer's Dementia **19**, e072907 (2023)
5. Khan, S.S., et al.: Unsupervised deep learning to detect agitation from videos in people with dementia. IEEE Access **10**, 10349–10358 (2022). https://doi.org/10.1109/ACCESS.2022.3143990
6. Mishra, P.K.: Automatic detection of behaviours of risk in people with dementia using unsupervised deep learning. In: Canadian AI (2022)
7. Gautam, C., Tiwari, A., Mishra, P.K., Suresh, S., Iosifidis, A., Tanveer, M.: Graph-embedded multi-layer kernel ridge regression for one-class classification. Cogn. Comput. **13**, 552–569 (2021)
8. Mishra, P.K., Gautam, C., Tiwari, A.: Minimum variance embedded auto-associative kernel extreme learning machine for one-class classification. Neural Comput. Appl. **33**(19), 12973–12987 (2021). https://doi.org/10.1007/s00521-021-05905-y
9. Mishra, P.K., Mihailidis, A., Khan, S.S.: Skeletal video anomaly detection using deep learning: Survey, challenges, and future directions. IEEE Trans. Emerg. Topics Comput. Intell. **8**(2), 1073–1085 (2024). https://doi.org/10.1109/TETCI.2024.3358103
10. Mishra, P.K., et al.: Depth-weighted detection of behaviours of risk in people with dementia using cameras. arXiv preprint arXiv:2408.15519 (2024)
11. Khan, S.S., et al.: A novel multi-modal sensor dataset and benchmark to detect agitation in people living with dementia in a residential care setting. ACM Transactions on Computing for Healthcare (2024)
12. Fook, V.F.S., : Automated recognition of complex agitation behavior of dementia patients using video camera. In: 2007 9th International Conference on e-Health Networking, Application and Services, pp. 68–73. IEEE (2007)
13. Qiu, Q., et al.: Multimodal information fusion for automated recognition of complex agitation behaviors of dementia patients. In: 2007 10th International Conference on Information Fusion, pp. 1–8. IEEE (2007)
14. Chikhaoui, B., Ye, B., Mihailidis, A.: Ensemble learning-based algorithms for aggressive and agitated behavior recognition. In: Ubiquitous Computing and Ambient Intelligence, pp. 9–20. Springer International Publishing, Cham (2016)

15. Lu, Y., Yu, F., Reddy, M., Wang, Y.: Few-shot scene-adaptive anomaly detection. In: Vedaldi, A., Bischof, H., Brox, T., Frahm, J.-M. (eds.) ECCV 2020. LNCS, vol. 12350, pp. 125–141. Springer, Cham (2020). https://doi.org/10.1007/978-3-030-58558-7_8

16. Lv, H., Chen, C., Cui, Z., Xu, C., Li, Y., Yang, J.: Learning normal dynamics in videos with meta prototype network. In: Proceedings of the IEEE/CVF Conference on Computer Vision aDnd Pattern Recognition, pp. 15425–15434 (2021)

17. Georgescu, M.I., Ionescu, R.T., Khan, F.S., Popescu, M., Shah, M.: A background-agnostic framework with adversarial training for abnormal event detection in video. IEEE Trans. Pattern Anal. Mach. Intell. **44**(9), 4505–4523 (2021)

18. Chen, L.C., Zhu, Y., Papandreou, G., Schroff, F., Adam, H.: Encoder-decoder with atrous separable convolution for semantic image segmentation. In: Proceedings of the European Conference on Computer Vision (ECCV), pp. 801–818 (2018)

19. Khan, S.S., Mishra, P.K., Ye, B., Newman, K., Iaboni, A., Mihailidis, A.: Empirical thresholding on spatio-temporal autoencoders trained on surveillance videos in a dementia care unit. In: 2023 20th Conference on robots and vision (CRV), pp. 265–272. IEEE (2023)

20. Spelmen, V.S., Porkodi, R.: A review on handling imbalanced data. In: 2018 international conference on current trends towards converging technologies (ICCTCT), pp. 1–11. IEEE (2018)

21. Huang, N., et al.: Segment any motion in videos (2025). https://arxiv.org/abs/2503.22268

Enhancing Diabetic Foot Ulcer Assessment Through Fine-Tuned Vision-Language Models

Reza Basiri[1,2]([✉])[iD], Asad Ghaffar[2], Donya Ghiasi[3], Meaka Tesfaye Mekonnen[3], Milos R. Popovic[1,2][iD], and Shehroz S. Khan[1,2,4][iD]

[1] KITE Research Institute, University Health Network, Toronto, Canada
reza.basiri@mail.utoronto.ca
[2] Institute of Biomedical Engineering, University of Toronto, Toronto, Canada
[3] Department of Surgery, University of Calgary, Calgary, Canada
[4] College of Engineering and Technology, American University of the Middle East, Al Ahmadi, Kuwait

Abstract. Diabetic foot ulcers (DFUs) are a significant cause of lower limb amputations and hospitalizations, placing a substantial burden on patients and healthcare systems. Early assessments are critical for preventing complications, yet the shortage of specialists cannot meet the widespread prevalence of this condition. This paper explores augmenting the DFU clinical workflow by applying vision-language models for generating clinically relevant assessments of DFUs from images. Using the Wound-Ischemia-Foot Infection (WIfI) classification system as a structured framework, we assess the performance of LLaVA-Mistral models fine-tuned on annotated DFU datasets compared to LLaVA-Mistral and GPT-4o baselines. Our findings demonstrate that the initial fine-tuned LLaVA-Mistral model achieved on average 16% higher average accuracy in predicting WIfI elements from a DFU image. In terms of clinical narrative generation quality, the LLaVA-Mistral model demonstrated a 22% improvement in DFU-specific text coherence compared to the baseline model as measured by the dependency parse tree depth method. This research lays the groundwork for AI-assisted DFU assessment by creating publicly available annotations for DFU vision-language model developments and demonstrating the potential of fine-tuning to enhance clinical communication through improved classification accuracy.

Keywords: Diabetic Foot Ulcer · Vision Language Models · LLaVA-Mistral · GPT-4o · WIfI Assessment

1 Introduction

Diabetes mellitus presents a substantial burden to aging populations, with prevalence rates exceeding 25% among adults aged 65 and older [12].[1] Diabetic foot

---

[1] To ensure reproducibility, the annotation dataset and code are provided through the DFU2022_LLM GitHub repository. Link: https://github.com/rezabasiri/DFU2022_LLM.

© The Author(s), under exclusive license to Springer Nature Singapore Pte Ltd. 2025
S. S. Khan et al. (Eds.): IJCAI 2025, CCIS 2620, pp. 117–131, 2025.
https://doi.org/10.1007/978-981-95-0568-5_9

ulcers (DFUs)–full-thickness lesions below the ankle resulting from peripheral neuropathy, vascular insufficiency, and prolonged pressure–are directly linked to diabetes mellitus and represent the leading cause of lower limb amputations among diabetic patients [3,6,11,17,23]. This heightened vulnerability stems from age-related changes in skin integrity, decreased peripheral circulation, delayed wound healing, and impaired mobility that prevents timely self-examination [11]. Consequently, DFUs impose substantial quality of life risks and economic burdens on healthcare systems, with treatment costs in the United States alone estimated to exceed $80 billion annually [2].

Clinical evaluation of DFUs has been standardized through several assessment systems. The Wagner classification grades ulcers from 0 to 5 based on wound depth and infection presence [30], while the University of Texas system incorporates both depth and infection-ischemia status in a matrix format [19]. The Wound-Ischemia-Foot-Infection (WIfI) classification provides comprehensive assessment by independently scoring three key factors impacting amputation risk and healing potential [25]. These structured approaches enable standardized documentation and evidence-based treatment protocols.

Traditional DFU management includes wound debridement, offloading techniques, and local wound care guided by systematic classification [13]. Recent AI advances have enhanced aspects of this workflow [8,10], with AI-driven image analysis demonstrating high binary classification accuracy of 93.4% using fully convolutional networks [14]. While recent work has explored multi-level scoring systems [31], these approaches typically focus on single classification frameworks and fail to capture comprehensive multi-dimensional clinical information required for treatment planning, such as that provided by the WIfI system [26]. Existing automated approaches often necessitate subsequent manual review, limiting AI-assisted diagnostic efficiency in clinical settings.

Vision-language models offer a potential solution for bridging this gap between image classification and clinical decision-making by generating comprehensive and interpretative assessments from visual input. Despite their promise, these models remain largely unexplored in DFU applications. Current models such as GPT-4o [27], Qwen-VL [4] and LLaVA-Mistral [18,21] struggle with DFU identification due to insufficient domain-specific training data [5]. While the publicly available DFU image datasets, such as DFU2022 [32], provide valuable images for domain-specific segmentation model developments, they lack the structured textual annotations required for training vision-language models to generate clinically relevant assessments.

This paper addresses these limitations through two primary contributions. First, we have annotated a portion of the DFU2022 dataset with structured WIfI categories, creating a resource specifically designed for training and evaluating vision-language models in the DFU domain. Second, we demonstrate the effectiveness of this annotated dataset by fine-tuning LLaVA-Mistral models to improve WIfI assessment text generation, comparing their results against GPT-4o as a state-of-the-art baseline. The WIfI assessment system was selected for this work due to its comprehensive approach to wound evaluation and established prognostic value in clinical settings, providing a framework that aligns well with the analytical capabilities of vision-language models.

The performance of these models was evaluated using quantitative metrics for classification accuracy while also assessing their potential for generating clinical narratives. This approach demonstrates the potential for automated WIfI-based assessment, which could significantly reduce the dependency on manual image review while improving diagnostic support through specialized vision-language models that bridge the gap between image analysis and clinical decision-making.

2   Related Work

2.1   AI-Driven DFU Assessment

AI applications in DFU assessment have evolved significantly across multiple modalities. Zhang et al. [33] conducted a comprehensive review noting that convolutional neural networks dominated classification tasks. For classification, Goyal et al. [14] introduced DFUNet with parallel convolutions, while Alzubaidi et al. [1] improved accuracy with increased DNN width. Recent advances have begun addressing binary classification limitations. Wang et al. [31] developed a smart diabetic foot ulcer scoring system that automatically assigns graded severity scores, demonstrating multi-level automated assessment feasibility. However, these approaches typically focus on single classification systems rather than comprehensive multi-dimensional assessment provided by frameworks like WIfI.

Monteiro-Soares et al. [26] conducted a critical review of diabetic foot ulcer classification systems, analyzing clinical relevance and limitations of existing frameworks including Wagner, UT, and WIfI classifications. Their analysis highlighted the superior prognostic value of multi-dimensional systems like WIfI compared to single-parameter approaches, supporting our selection of the WIfI framework. The review emphasized that while various classification systems exist, standardization and automation remain significant clinical challenges.

2.2   Vision-Language Models in Medical Imaging

Vision-language models combine computer vision with natural language processing to generate image assessments. In the medical domain, these models have shown promise in tasks such as chest radiography interpretation and radiology report generation [16]. Models such as LLaVA-Med [20] have succeeded in various radiology benchmarks by replacing standard vision encoders with domain-specific components (e.g., BioCLIP [29]) and utilizing GPT-4-assisted labels for medical data.

Basiri et al. [5] conducted a comprehensive evaluation of several vision-language models for DFU image description, including GPT-4o, Qwen-VL, and various LLaVA configurations. Their findings indicated that GPT-4o demonstrated the strongest overall performance with average clinical ratings of 3.6/5 across all metrics evaluated by expert clinicians. Qwen-VL showed promising intermediate results (3.3/5), particularly in accurately identifying ulcer locations and providing comprehensive descriptions of wound characteristics. LLaVA-Mistral achieved moderate performance (2.3/5) but struggled with reliably identifying DFUs, often referring vaguely to "skin conditions" or incorrectly describing foot anatomy. The study also noted that all models, including the best-performing GPT-4o, demonstrated substantial deviation from expert-level assessment, highlighting significant opportunities for improvement through domain-specific adaptation and training.

These findings suggest a significant opportunity for improvement through specialized training approaches targeted at structured clinical assessment frameworks such as WIfI, which remains unexplored in current literature. Our work aims to address this gap by first establishing a publicly available vision-language model compatible dataset and second enhancing model performance through targeted fine-tuning on these DFU annotations.

3   Methodology

3.1   Overview

This study evaluates the effectiveness of fine-tuning vision-language models for classifying DFU images according to the WIfI framework. We employ the WIfI elements, Wound-Ischemia-Foot-Infection, into a structured framework to assess model performance. This approach enables quantitative evaluation of the models' ability to identify key clinical characteristics in DFU images. We additionally evaluated the quality of the DFU image narrations using the vision-language models.

3.2   Dataset Preparation

For the dataset preparation, 1,000 DFU images were randomly selected from the Manchester 2022DFU dataset [32]. Building upon prior work establishing cross-modal transitions between different DFU representations [7,9], a specialized annotation interface ("WoundVista Annotator") was developed and implemented as illustrated in Fig. 1. This custom tool facilitated the collection of structured WIfI annotations across seven clinically relevant categories: Wound, Ischemia, Foot Infection, Surrounding Skin Condition, Wound Size, Ulcer Location, and Swelling. The annotation protocol was designed to capture standardized clinical parameters while minimizing inter-annotator variability through structured data entry fields.

Two domain expert clinicians independently annotated the image dataset in parallel sessions. Inter-annotator reliability was evaluated using Cohen's Kappa coefficient [24] for categorical variables and percentage agreement for both primary WIfI categories and their respective subcategories. To ensure data quality, only annotations achieving consensus between both clinicians were incorporated into the final dataset. Consequently, the comprehensive DFU characterization for each image comprised a variable subset of WIfI categories, with inclusion determined by the strength of inter-annotator agreement. This consensus-driven

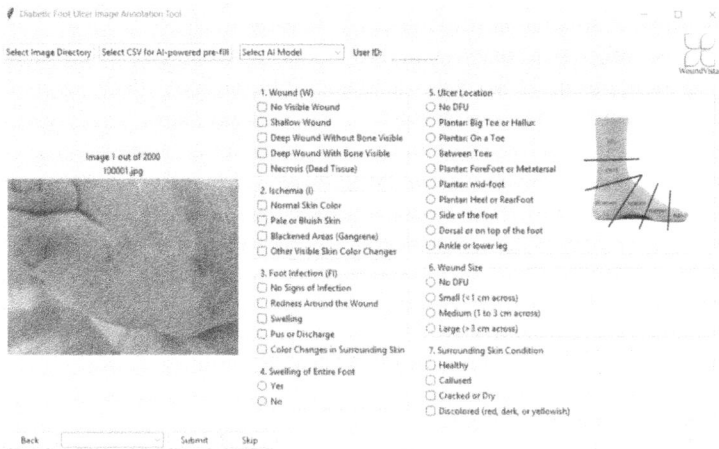

Fig. 1. The WoundVista Annotator Tool. This interface was used by clinicians to collect structured WIfI labels from DFU images.

approach prioritized annotation reliability at the expense of categorical completeness, resulting in heterogeneous but high-confidence clinical descriptions across the dataset.

The dataset curated using the WoundVista Annotator, contained 21 categories in total, with 19 binary categories and two non-binary categories (Ulcer Location and Wound Size) as shown in Fig. 1. The binary categories covered various aspects of diabetic foot ulcer assessment, including: I. Wound characteristics (presence, depth, and necrosis), II. Infection indicators (swelling, discharge, infectious discolorations), III. Surrounding skin conditions (healthy, callused, cracked, discolorations), IV. Ischemia indicators (gangrene, ischemic discolorations), and V. Additional indicator for swelling of the entire foot. The non-binary categories provided more nuanced information, including ulcer Location (10 possible locations) and wound Size (4 size categories). For each image, these structured annotations were transformed into natural language descriptions using a template-based approach that converted binary and categorical labels into clinically coherent narratives while maintaining professional terminology. The prompt templates were designed to elicit structured responses corresponding to each WIfI category.

For DFU descriptive fine-tuning, the annotated data was structured in JSON format adhering to a standardized chat template architecture to facilitate fine-tuning procedures. Each entry contained structured elements including:

1. Image identifier: A unique filename reference (e.g., 100045.jpg)
2. Conversational exchange: A simulated dialogue comprising:
   - Human query component: A standardized prompt requesting WIfI assessment interpretation (e.g., Describe the diabetic foot ulcer in this image for WIfI assessment.)

- Assistant response component: A comprehensive clinical description incorporating consensus-validated WIfI parameters (e.g., The image shows a medium (1 to 3 cm across) diabetic foot ulcer located on the plantar forefoot (metatarsal area). The image shows a shallow diabetic foot ulcer with necrotic tissue (dead tissue). Signs of infection are evident, including redness around the wound and swelling around the wound area. The skin surrounding the ulcer appears discolored (red, dark, or yellowish). The image indicates ischemia characterized by pale or bluish skin.)

The dataset was split into training (70%), validation (15%), and test (15%) sets, based on unique images to avoid data leakages across splits.

3.3   Model Architecture and Implementation

Based on the comparative evaluation of vision-language models for DFU assessment [5], LLaVA-Mistral was selected as the foundation architecture for this investigation. This selection was predicated on multiple methodological considerations: (1) the computational efficiency of the model aligned with available research infrastructure constraints; (2) its open-source implementation facilitated reproducibility and methodological transparency; and (3) the modular architecture and compatibility with different language components supported experimental variations. The implementation utilized 'llava-v1.6-mistral-7b', which integrates the Contrastive Language-Image Pre-training (CLIP) vision encoder ('openai/clip-vit-large-patch14-336') [28] with the Mistral 7B language model through a projection layer that facilitates cross-modal information transfer. This architectural configuration enables the processing of visual inputs from DFU images and the generation of structured clinical narratives incorporating WIfI classification parameters, establishing a technical foundation for domain-specific adaptation through supervised fine-tuning procedures.

The experiments were conducted on the Digital Research Alliance of Canada, Cedar cluster, using V100L GPU nodes. The LLaVA-Mistral model and CLIP vision tower were downloaded from the Hugging-Face repository and stored locally.

3.4   Experimental Setup

For comparison, we established GPT-4o (gpt-4o-2024-08-06) as our performance baseline, representing state-of-the-art in multimodal language models. We also evaluated the performance of the unmodified LLaVA-Mistral model to measure the impact of fine-tuning.

Fine-tuning was performed using Low-Rank Adaptation (LoRA) [15] to optimize resource efficiency. This technique introduces trainable low-rank matrices into specific model layers, enabling adaptation without modifying the original pre-trained weights, thus significantly reducing computational requirements.

Hyperparameter selection was performed through grid search on the validation set, varying batch size (8, 16) and learning rate (1e-6, 1e-4).

For inference and using the test dataset, the standardized prompts guided the models in providing specific textual outputs for each WIfI category.

3.5   Evaluation Metrics

Performance was measured under two approaches: 1. For each WIfI category, 2. Overall quality of the WIfI text description. For the first approach, the presence or absence of each WIfI condition in the image was inquired during the inference, limiting the response to a binary yes or no. This enabled us to statistically measure the accuracy, the percentage of correct classifications across the WIfI components, and F1-score, the harmonic mean of precision and recall, of the models' classification in a binary format. Statistical significance for these two measures was assessed using t-tests ($p < 0.05$).

For the second approach and to assess readability, we apply a syntactic complexity measure based on dependency parse tree depth (DPTP), which quantifies the hierarchical structure of sentences. The dependency tree, derived from a syntactic parser, represents the grammatical relationships between words, and its depth reflects the longest path from the root to any leaf node. Prior work in computational linguistics has used parse tree depth as an indicator of sentence complexity, with typical well-formed sentences ranging between depths of 10âĂŞ25, while unusually deep trees (e.g., >40) may signal fragmented lists or poor cohesion. Although not a standard text generation evaluation metric, dependency tree depth has been employed in linguistic and readability analyses and serves here to provide structural insights into the coherence of generated clinical narratives [22].

4   Results

4.1   Dataset Annotation Analysis

Inter-annotator agreement varied considerably across WIfI categories, as shown in Fig. 2 and measured by Cohen's Kappa score. While the majority of the categories show some level of agreement, three categories relevant to the complete absence of infection, skin dryness surrounding the ulcer and gangrene colour exhibited no agreement among the two annotators, and were thus excluded from the rest of the analysis.

Despite this variability, percentage agreement was generally high across categories, suggesting that many images had clear, unambiguous classifications.

4.2   WIfI Category Classification Performance

Table 1 summarizes the performance of different models on WIfI classification tasks. The fine-tuned LLaVA-Mistral model achieved the highest accuracy among vision-language models (76.41% $\pm$ 21.03%), outperforming the GPT-4o baseline (65.09% $\pm$ 14.57%). However, this difference was not statistically significant (p = 0.05).

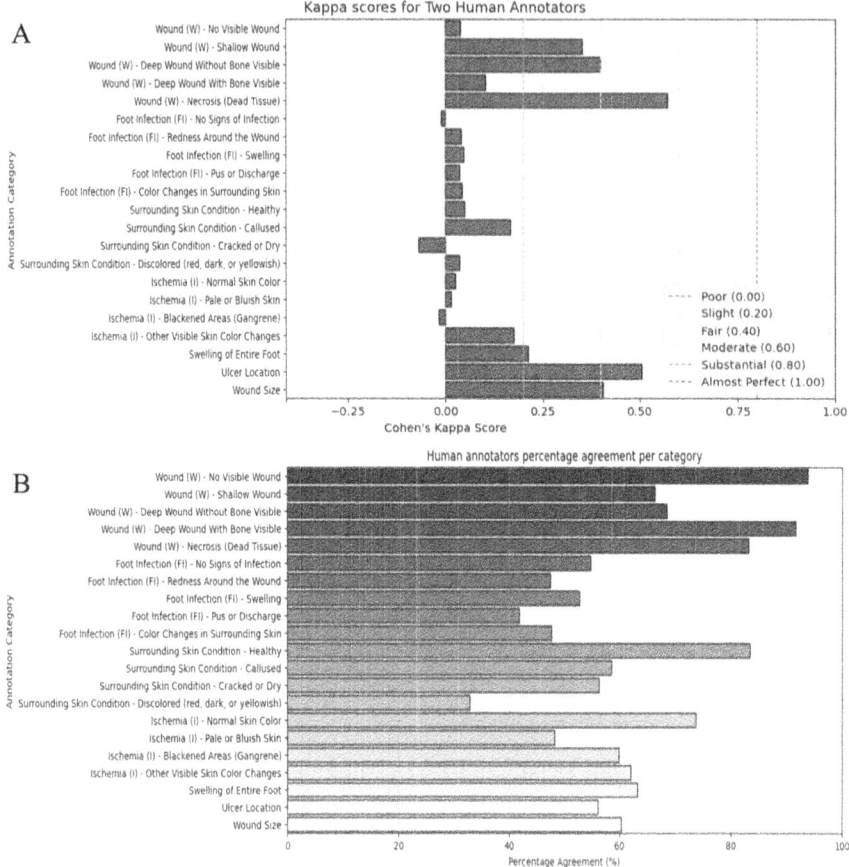

Fig. 2. Annotation Evaluations. The (A) Cohen's Kappa scores and (B) percentage agreement for the two clinician annotators across WIfI categories show variable levels of agreement.

4.3 Misclassification Pattern Analysis

Examination of the classification errors reveals distinct patterns across models and WIfI categories. The fine-tuned LLaVA-Mistral model demonstrated superior performance in visually distinct categories such as "Wound Size" (accuracy: 89.2%) and "Ulcer Location" (accuracy: 85.6%), where clear anatomical landmarks facilitate accurate assessment. However, performance degraded in subjective categories requiring clinical inference, particularly "Ischemia - Pale/Bluish Skin" (accuracy: 62.3%) and "Infection - Redness" (accuracy: 58.7%).

GPT-4o exhibited more consistent performance across categories, with standard deviation of 14.57% compared to 21.03% for fine-tuned LLaVA-Mistral. This consistency likely reflects GPT-4o's extensive pre-training on medical content, enabling better generalization to challenging cases. Notably, GPT-4o out-

Table 1. Average accuracy and F1-Score across WIfI categories

Model	Accuracy (%)	F1-Score
GPT-4o Baseline	65.09 ± 14.57	0.51 ± 0.09
LLaVA-Mistral Baseline	60.24 ± 32.96	0.39 ± 0.21
LLaVA-Mistral Fine-Tuned	76.41 ± 21.03	0.44 ± 0.17

performed both LLaVA variants in infection-related categories, suggesting that clinical reasoning capabilities benefit from broader medical knowledge rather than domain-specific fine-tuning alone.

The high variance in fine-tuned LLaVA-Mistral performance correlates with inter-annotator agreement patterns (Fig. 2), indicating that model performance is constrained by annotation quality. Categories with Cohen's kappa values below 0.4 consistently showed reduced model accuracy across all evaluated systems.

The high standard deviations in the fine-tuned LLaVA-Mistral's performance reflect inconsistent accuracy across different WIfI annotations. Figure 3 illustrates this variability, with the model excelling in categories with consistent annotation agreements, such as 'Wound - No Visible Wound' or Ischemia categories. However, LLaVA Fine-Tuned also demonstrated lower F1 scores in several categories despite higher accuracy, suggesting performance issues stemming from data imbalance. This imbalance is particularly evident in categories with predominantly negative examples, where the model achieves high accuracy by frequently predicting the majority class but fails to correctly identify the rarer positive cases. In contrast, GPT-4o exhibited more consistent performance across categories, as shown in Fig. 3, with lower overall accuracy but less variation between categories and more balanced precision-recall trade-offs, indicating greater robustness to class imbalance issues.

4.4 Clinical Narration Quality

When leveraging WIfI classifications to inform clinical narrative generation, we observed improvements in both semantic relevance and linguistic structure. Table 2 presents the DPTD metric for the baseline LLaVA-Mistral and when using WIfI classification as conditioning information.

The WIfI-guided approach demonstrated a 22.2% improvement in DFU-specific linguistic structure compared to the baseline as measured by DPTD, where lower values indicate more efficient linguistic structure. This indicates that the structured WIfI classifications effectively enhanced the clinical relevance of the generated narratives without compromising linguistic coherence.

5   Discussion

Our study demonstrates that fine-tuning vision-language models on domain-specific annotations significantly enhances their performance in specialized medical assessment tasks. The implementation of the WIfI framework provided a

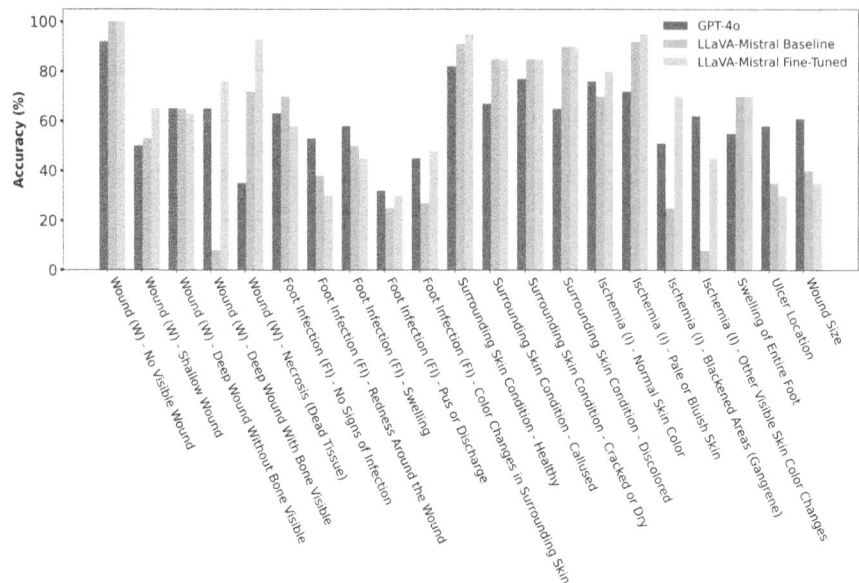

Fig. 3. Model Accuracy Comparison. Accuracy for binary prediction of WIfI categories done by each model. Overall, the LLaVA-Mistra fine-tuned model shows increased performance compared to the other two models.

Table 2. DPTD values for clinical narrative generation

Model	DPTD (Mean ± SD)
LLaVA-Mistral Baseline	26.95 ± 4.93
LLaVA-Mistral Fine-Tuned	20.97 ± 3.96*

*Decreased values indicate improved linguistic efficiency and cohesion.

structured approach to evaluate the capabilities of these models in DFU assessment, revealing both promising advances and areas requiring further development.

5.1 Dataset Annotation and Inter-Annotator Agreement

The variable inter-annotator agreement across WIfI categories highlights the inherent challenges in establishing reliable ground truth data for DFU assessment. The highest agreement was observed in objective categories such as "Wound Size" and "Ulcer Location," while more subjective assessments like "Ischemia" showed moderate agreement. This pattern suggests that visual characteristics that can be objectively quantified yield more consistent annotations than those requiring clinical inference or interpretation.

Despite these annotation challenges, the consensus-based approach generated sufficient high-quality training data to enable effective model fine-tuning. The correlation between inter-annotator agreement and model performance suggests that future work should prioritize annotation protocols with clearer operational definitions for subjective categories, potentially incorporating additional expert reviewers to resolve discrepancies. This finding aligns with previous research demonstrating that annotation consistency directly impacts model performance in medical image analysis [33].

5.2   Model Performance Analysis

The fine-tuned LLaVA-Mistral model demonstrated a 16.2% improvement in WIfI classification accuracy compared to the baseline version, achieving performance levels comparable to or exceeding the GPT-4o baseline. This improvement is particularly noteworthy given the resource efficiency of our approach, which utilized LoRA fine-tuning rather than end-to-end retraining of the entire model architecture.

The performance variability across different WIfI categories reveals an important pattern: the fine-tuned model excelled in categories with consistent annotations and clear visual manifestations, while struggling with more nuanced clinical assessments requiring deeper domain knowledge. This suggests that LoRA fine-tuning primarily enhances the vision-language alignment for clearly defined visual features without necessarily improving the underlying clinical reasoning capabilities of the model.

GPT-4o's more consistent performance across all categories, despite a lower overall accuracy, indicates its stronger generalization capabilities and more robust understanding of medical concepts. This pattern likely stems from GPT-4o's extensive pre-training on diverse medical content, compared to the more focused but limited fine-tuning approach applied to LLaVA-Mistral. Similar observations have been noted in prior comparative studies of large foundation models versus domain-adapted smaller models [5].

5.3   Clinical Narrative Generation Quality

The most striking improvement was observed in the quality of clinical narratives generated by the fine-tuned model. The 22.2% improvement in DFU-specific coherence narration as exhibited by the DPTD measure suggests that the fine-tuning process effectively calibrated the model's vocabulary toward clinically appropriate terminology when presented DFU images while capturing the complexity of WIfI clinical assessments.

These improvements have important clinical implications. Generated narratives that balance medical accuracy with readability can facilitate communication between specialists and non-specialists in multidisciplinary care settings. The structured WIfI framework provides an organizing principle for these narratives, ensuring comprehensive coverage of clinically relevant features while maintaining compatibility with standardized documentation systems.

However, qualitative analysis of the generated narratives revealed persistent limitations. Even the fine-tuned model occasionally produced generic or imprecise descriptions for complex cases, particularly when presented with unusual presentations or comorbidities not well-represented in the training data. This finding is consistent with previous evaluations of vision-language models in specialized medical domains [20], highlighting the continued importance of human oversight in clinical applications.

5.4   Limitations and Future Directions

Several limitations should be acknowledged when interpreting our results. First, the annotation dataset, while substantial, represents a limited cross-section of possible DFU presentations and may not fully capture the clinical diversity encountered in real-world settings. Second, our fine-tuning approach using LoRA primarily adapts the language model components without end-to-end optimization of the vision-language alignment, potentially limiting improvements in visual feature extraction.

Future work should address these limitations through multiple avenues. Expanding the annotation dataset with a broader range of DFU presentations and implementing more structured annotator training protocols would improve the foundation for model learning. Exploring alternative fine-tuning approaches that optimize both vision and language components simultaneously could potentially resolve the inconsistent performance across WIfI categories.

Clinical validation represents another critical direction for future research. Formal assessment of generated narratives by a broader panel of clinicians would provide valuable insights into the real-world utility of these models. Developing specialized evaluation metrics beyond classification accuracy could better capture clinically relevant aspects of model performance. Additionally, integrating LLaVA-Med with our fine-tuning approach could leverage its pre-existing medical knowledge to enhance performance on edge cases and rare presentations.

Beyond technical improvements, future research should explore integrating these models into telehealth workflows, where automated assessment could support remote clinicians in triaging and monitoring DFUs. This application has particular potential for underserved or resource-limited settings, where access to specialized wound care expertise may be limited. By providing consistent, structured assessments based on the clinically validated WIfI framework, these models could help standardize care and improve outcomes while reducing the burden on healthcare systems.

6   Conclusion

This study demonstrates the potential of fine-tuned vision-language models for enhancing DFU assessment through the structured WIfI classification framework. Our findings make several important contributions to the field of medical

AI and DFU management. Domain-specific fine-tuning of LLaVA-Mistral produced a 16.2% improvement in classification accuracy over the baseline model, demonstrating that efficient fine-tuning approaches can enhance performance for specialized medical tasks. The fine-tuned model also generated substantially improved clinical narratives, with a 22.2% enhancement in DFU-specific linguistic structure, supporting more effective clinical communication. Our comparison with GPT-4o revealed complementary strengths: while fine-tuned LLaVA-Mistral achieved higher overall accuracy, GPT-4o showed more consistent cross-category performance, suggesting potential value in hybrid approaches. We have made our dataset and code publicly available through the "DFU2022_LLM" GitHub repository to support reproducibility and future research. While these advances are promising, current models should be viewed as supportive tools that augment rather than replace clinical expertise, particularly for complex cases requiring nuanced judgment. Nevertheless, this work establishes a pathway toward AI-assisted wound care that could improve standardization, efficiency, and access to expertise for patients with diabetic foot complications.

Acknowledgments. This research was supported by computational resources provided by the Digital Research Alliance of Canada.

Disclosure of Interests. The authors have no competing interests to declare that are relevant to the content of this article.

References

1. Alzubaidi, L., Fadhel, M.A., Oleiwi, S.R., Al-Shamma, O., Zhang, J.: DFU_QUTNet: diabetic foot ulcer classification using novel deep convolutional neural network. Multimed. Tools Appl. **79**(21), 15655–15677 (2020). https://doi.org/10.1007/s11042-019-7719-y
2. Armstrong, D.G., Swerdlow, M.A., Armstrong, A.A., Conte, M.S., Padula, W.V., Bus, S.A.: Five year mortality and direct costs of care for people with diabetic foot complications are comparable to cancer. J. Foot Ankle Res. **13**, 1–4 (2020) https://doi.org/10.1186/s13047-020-00383-2
3. Armstrong, D.G., Tan, T.W., Boulton, A.J., Bus, S.A.: Diabetic foot ulcers: a review. JAMA **330**(1), 62–75 (2023). https://doi.org/10.1001/jama.2023.10578
4. Bai, J., et al.: Qwen-VL: A versatile vision-language model for understanding, localization, text reading, and beyond. arXiv preprint arXiv:2308.12976 (2023)
5. Basiri, R., Abedi, A., Nguyen, C., Popovic, M.R., Khan, S.S.: UlcerGPT: A multimodal approach leveraging large language and vision models for diabetic foot ulcer image transcription. In: Pattern Recognition. ICPR 2024 International Workshops and Challenges. Lecture Notes in Computer Science, vol. 15618, pp. 239–253. Springer, Cham (2025).https://doi.org/10.1007/978-3-031-88220-3_16
6. Basiri, R., Haverstock, B.D., Petrasek, P.F., Manji, K.: Reduction in diabetes-related major amputation rates after implementation of a multidisciplinary model: an evaluation in Alberta, Canada. J. American Podiatric Med. Assoc. **111**(4) (2021)https://doi.org/10.7547/20-022

7. Basiri, R., Manji, K., Francois, H., Poonja, A., Popovic, M.R., Khan, S.S.: Synthesizing diabetic foot ulcer images with diffusion model. In: Machine Learning and Principles and Practice of Knowledge Discovery in Databases. ECML PKDD 2023. Communications in Computer and Information Science, vol. 2136. Springer, Cham (2025). https://doi.org/10.1007/978-3-031-74640-6_28

8. Basiri, R., et al.: Protocol for metadata and image collection at diabetic foot ulcer clinics: enabling research in wound analytics and deep learning. Biomed. Eng. Online **23**(1), 12 (2024). https://doi.org/10.1186/s12938-024-01198-4

9. Basiri, R., Mestral, C.d., Popovic, M.R., Khan, S.S.: Accessible healing phase classification of diabetic foot ulcer (2025), preprint

10. Basiri, R., Popovic, M.R., Khan, S.S.: Domain-specific deep learning feature extractor for diabetic foot ulcer detection. In: 2022 IEEE International Conference on Data Mining Workshops (ICDMW), pp. 1–5. IEEE (2022). https://doi.org/10.1109/ICDMW58026.2022.00006

11. Brem, H.A., et al.: Healing of elderly patients with diabetic foot ulcers, venous stasis ulcers, and pressure ulcers. Surg. Technol. Int. **11**, 161–167 (2001)

12. Centers for Disease Control and Prevention: National diabetes fact sheet: national estimates and general information on diabetes and prediabetes in the United States (2011)

13. Everett, E., Mathioudakis, N.: Update on management of diabetic foot ulcers. Ann. N. Y. Acad. Sci. **1411**(1), 153–165 (2018). https://doi.org/10.1111/nyas.13569

14. Goyal, M., Reeves, N.D., Davison, A.K., Rajbhandari, S., Spragg, J., Yap, M.H.: DFUNet: convolutional neural networks for diabetic foot ulcer classification. IEEE Trans. Emerg. Topics Comput. Intell. **4**(5), 728–739 (2020). https://doi.org/10.1109/TETCI.2018.2866254

15. Hu, E.J., et al.: LoRA: Low-rank adaptation of large language models. In: International Conference on Learning Representations (2022)

16. Hu, M., Qian, J., Pan, S., Li, Y., Qiu, R.L., Yang, X.: Advancing medical imaging with language models: featuring a spotlight on ChatGPT. Phys. Med. Biol. **69**(10), 10TR01 (2024).https://doi.org/10.1088/1361-6560/ad4766

17. Jaul, E., Barron, J., Rosenzweig, J.P., Menczel, J.: An overview of co-morbidities and the development of pressure ulcers among older adults. BMC Geriatrics **18**, 1–11 (2018) https://doi.org/10.1186/s12877-018-0997-7

18. Jiang, A.Q., et al.: Mistral 7B. arXiv preprint arXiv:2310.06825 (2023) https://doi.org/10.48550/arXiv.2310.06825

19. Lavery, L.A., Armstrong, D.G., Harkless, L.B.: Classification of diabetic foot wounds. J. Foot Ankle Surg. **35**(6), 528–531 (1996). https://doi.org/10.1016/S1067-2516(96)80125-6

20. Li, C., Wong, C., Zhang, S., et al.: LLaVA-Med: Training a large language-and-vision assistant for biomedicine in one day. In: Advances in Neural Information Processing Systems, vol. 36 (2023)

21. Liu, H., Li, C., Wu, Q., Lee, Y.J.: Visual instruction tuning. In: Advances in Neural Information Processing Systems. vol. 36, pp. 34892–34916 (2023)

22. Lu, X.: Automatic analysis of syntactic complexity in second language writing. Int. J. Corpus Linguist. **15**(4), 474–496 (2010). https://doi.org/10.1075/ijcl.15.4.02lu

23. Manji, A., Basiri, R., Harton, F., Rommens, K., Manji, K.: Effectiveness of a multidisciplinary limb preservation program in reducing regional hospitalization rates for patients with diabetes-related foot complications. Int. J. Low. Extrem. Wounds **24**(1), 117–123 (2025). https://doi.org/10.1177/15347346231221052

24. McHugh, M.L.: Interrater reliability: the kappa statistic. Biochemia Medica **22**(3), 276–282 (2012). https://doi.org/10.11613/BM.2012.031

25. Mills, J.L., et al.: The society for vascular surgery lower extremity threatened limb classification system: risk stratification based on wound, ischemia, and foot infection (WIfI). J. Vasc. Surg. **59**(1), 220–234 (2014). https://doi.org/10.1016/j.jvs.2013.08.003

26. Monteiro-Soares, M., et al.: Diabetic foot ulcer classifications: a critical review. Diabetes Metab. Res. Rev. **36**(S1), e3272 (2020). https://doi.org/10.1002/dmrr.3272

27. OpenAI: Hello GPT-4o (2024). https://openai.com/index/hello-gpt-4o/. Accessed 26 Sept 2024

28. Radford, A., et al.: Learning transferable visual models from natural language supervision. In: International Conference on Machine Learning, pp. 8748–8763. PMLR (2021)

29. Stevens, S., et al.: BioCLIP: A vision foundation model for the tree of life. In: Proceedings of the IEEE/CVF Conference on Computer Vision and Pattern Recognition, pp. 19412–19424 (2024)

30. Wagner, F.W.: The dysvascular foot: a system for diagnosis and treatment. Foot Ankle Int. **2**(2), 64–122 (1981). https://doi.org/10.1177/107110078100200202

31. Wang, Z., et al.: Smart diabetic foot ulcer scoring system. Sci. Rep. **14**(1), 11588 (2024). https://doi.org/10.1038/s41598-024-62282-4

32. Yap, M.H., et al.: Diabetic foot ulcers segmentation challenge report: Benchmark and analysis. Med. Image Analy. **94**, 103153 (2024).https://doi.org/10.1016/j.media.2024.103153

33. Zhang, J., Qiu, Y., Peng, L., Zhou, Q., Wang, Z., Qi, M.: A comprehensive review of methods based on deep learning for diabetes-related foot ulcers. Front. Endocrinol. **13**, 945020 (2022). https://doi.org/10.3389/fendo.2022.945020

Brain Imaging and Tissue Segmentation from Paired CT-MRI Labels for Cognitively Normal and Dementia Cohort

Vidya Somashekarappa[1,2(✉)] ⓘ, Meera Srikrishna[1,2] ⓘ, Silke Kern[2,3,4] ⓘ,
Joyce Chong Ruifen[4,5,6] ⓘ, Eric Westman[7] ⓘ, Christopher Chen[5,6] ⓘ,
Ingmar Skoog[7] ⓘ, Jakob Seidlitz[8] ⓘ, and Michael Schöll[1,2,9] ⓘ

[1] Wallenberg Centre for Molecular and Translational Medicine, University of Gothenburg, Gothenburg, Sweden
{vidya.somashekarappa,meera.srikrishna}@gu.se, michael.scholl@neuro.gu.se
[2] Department of Psychiatry and Neurochemistry, Institute of Physiology and Neuroscience, Gothenburg, Sweden
silke.kern@neuro.gu.se
[3] Neuropsychiatric Epidemiology, Centre for Ageing and Health (AgeCap), University of Gothenburg, Gothenburg, Sweden
[4] Department of Neuropsychiatry, Sahlgrenska University Hospital, Gothenburg, Sweden
phcjcr@nus.edu.sg
[5] Memory Aging and Cognition Centre, Singapore, Singapore
cplhchen@yahoo.com.sg
[6] Department of Pharmacology, National University of Singapore, Singapore, Singapore
[7] Division of Clinical Geriatrics, Karolinska Institutet, Solna, Sweden
eric.westman@ki.se, ingmar.skoog@neuro.gu.se
[8] Lifespan Brain Institute, The Children's Hospital of Philadelphia and Penn Medicine, Philadelphia, USA
jakob.seidlitz@pennmedicine.upenn.edu
[9] Dementia Research Centre, University College London, London, UK

Abstract. Brain tissue segmentation plays a fundamental role in clinical and research settings, creating detailed brain atlases, diagnosing neurological disorders such as dementia, and planning surgical interventions. Although magnetic resonance imaging is the preferred modality for its superior soft tissue contrast, CT serves as an accessible alternative to imaging in clinical routine and emergency situations. In this study, paired CT-MRI data sets from the Gothenburg H70 Birth Cohort ($N=733$) and the National University Hospital Memory Clinic Cohort, Singapore (NUS dementia cohort $N=210$) were used to train and evaluate advanced segmentation frameworks–nnUNet and MedNeXt–on CT brain segmentation guided by MRI-derived labels. In addition, the best performing models trained on Axial CT were selected to evaluate the sagittal and coronal CT. The 3D nnU-Net achieved average Dice Similarity Coefficients (DSCs) of 0.82, 0.72, and 0.76 for axial, coronal, and sagittal datasets,

© The Author(s), under exclusive license to Springer Nature Singapore Pte Ltd. 2025
S. S. Khan et al. (Eds.): IJCAI 2025, CCIS 2620, pp. 132–146, 2025.
https://doi.org/10.1007/978-981-95-0568-5_10

respectively, while MedNeXt exhibited slightly better performance with DSCs of 0.83, 0.73, and 0.78. MedNeXt demonstrated improved volumetric similarity across orientations, particularly in axial datasets, with scores ranging from 0.842 (CSF, sagittal) to 0.992 (WM, axial). For generalizability of the cohort to dementia patients, MedNeXt outperformed nnU-Net, achieving an average DSC and volumetric similarity of 0.73 and 0.912, compared to 0.70 and 0.854 for nnU-Net. Extended training (1000 epochs) improved the performance of nnUnet, while MedNeXt demonstrated superior scalability with increasing kernel sizes, requiring significantly longer training times (up to 288 h for large models). Thus, we demonstrate the higher training efficiency of nnUNet (requiring lower resource usage) and inference times, hence suitable for environments with limited resources, whereas MedNeXt consistently performs better and excels in multimodal imaging scenarios while maintaining clinically relevant accuracy across diverse datasets. Collectively, these findings validate automated brain tissue segmentation models for CT using paired MRI.

Keywords: Brain segmentation · Deep Learning · Brain Imaging · Dementia

1   Introduction

1.1   Brain Segmentation

Brain segmentation is an essential technique in clinical and research settings, with applications ranging from the creation of detailed brain atlases, the diagnosis and monitoring of neurological disorders such as dementia, multiple sclerosis, and epilepsy, to planning surgical interventions and evaluating tumor growth or response to therapy [4,5]. In patients with dementia, brain segmentation helps identify patterns of atrophy associated with disease progression [11], while in hydrocephalus it helps diagnose ventricular enlargement. Magnetic Resonance Imaging (MRI) is the modality of choice for brain segmentation due to its superior soft tissue contrast and resolution, allowing detailed visualization of brain structures [3,16]. However, computed tomography (CT) serves as an accessible and economic alternative, especially in routine clinical settings and in emergencies where rapid imaging is crucial, such as in acute head injuries or suspected stroke [1]. The complementary roles of MRI and CT underscore the versatility of imaging technologies in advancing brain segmentation and clinical care.

1.2   Deep Learning and SOTA in MRI and CT

Deep learning has revolutionized medical imaging, achieving excellent segmentation performance in MRI and CT analysis through architectures like U-Net and its variants, as well as transformer-based models. U-Net [13], originally introduced for biomedical image segmentation, has inspired numerous variants such as ResU-Net [12], Attention U-Net [7], and Dense U-Net   [19], which improve segmentation precision by integrating spatial and contextual information. Transformer-based models, with their ability to capture long-range dependencies, are useful for applications such as lesion detection, tissue classification, and volumetric segmentation, often outperforming traditional convolutional approaches, especially when combined with convolutional layers for hybrid architectures [15]. Studies have used deep learning in CT imaging, particularly in optimizing low-dose CT reconstructions using denoising autoencoders and GAN [18], as well as detecting abnormalities such as intracranial hemorrhages [10]. These advancements highlight the transformative impact of deep learning, enabling enhanced diagnostic accuracy and efficiency in medical imaging.

1.3   nnUNet and MedNeXt

Building on these advancements, frameworks like nnU-Net and MedNeXt have emerged as essential tools addressing the need for adaptable, efficient, and high-performing architectures in medical imaging. nnU-Net, recognized as a universal framework for biomedical image segmentation, enables efficient workflows by automatically adapting preprocessing, architecture, and training configurations to the specific needs of a dataset, minimizing manual intervention and improving generalization [6]. MedNeXt complements this adaptability by introducing a hybrid architecture ConvNeXt that integrates convolutional and transformer-based approaches, achieving high scalability and precision in tasks such as MRI and CT image segmentation [14]. Its design allows the model to scale the network width and learn long-range dependencies through large kernels. This architecture also addresses overfitting, a common issue in training large networks on limited datasets. Initial adaptations of ConvNeXt for medical image segmentation have achieved notable success. For example, a large kernel 3D-UNet has demonstrated improved segmentation performance in tasks such as organ and brain tumor segmentation by decomposing kernels into depth-wise and depth-wise dilated components [9]. Similarly, 3D-UX-Net replaced Swin Transformer blocks in SwinUNETR with ConvNeXt components, resulting in enhanced performance in multiple segmentation tasks [8]. Together, nnU-Net and MedNeXt exemplify the evolution of deep learning in medical imaging, bridging the gap between traditional state-of-the-art architectures and the next generation of domain-agnostic frameworks.

1.4   Research Gap and Motivation

Previously, research has demonstrated the utility of 2D U-Nets to segment CT brain scans but also highlighted the limitations in capturing the full spatial context and volumetric intricacies of brain tissues [17]. Furthermore, limited studies have systematically compared advanced frameworks such as nnU-Net and MedNeXt in this unique paired imaging scenario. This study aims to address these gaps by evaluating the performance and adaptability of these newer frameworks for CT segmentation tasks guided by MRI labels, thus exploring their potential to advance multimodal imaging analysis.

2   Methodology

2.1   Dataset

Gothenburg H70 Birth Cohort. The Gothenburg H70 Birth Cohort represents a series of ongoing epidemiological studies initiated in 1971, examining large representative samples of 70-year-old residents of the Gothenburg area, Sweden. Data from the 2014–2016 cohort, consisting of 1,203 participants in total, were used in this study. Cranial CT imaging was available for 917 participants, most of whom (99%; n = 904) were cognitively normal. Of these, 744 individuals (79%) also underwent MRI within 24 h of their CT scans, which provided us with CT-MRI acquisitions minimizing anatomical discrepancies due to temporal changes. Imaging procedures were performed at Aleris Röntgen Annedal in Göteborg, Sweden, using a 64-slice Ingenuity CT scanner (Philips Medical Systems, Best, The Netherlands) for CT acquisitions and a 3 T Achieva MRI system (Philips Medical Systems) for MRI scans. Informed consent was obtained from all participants and safety assessments were performed through interviews before imaging. In addition to the imaging data, demographic information, medical history and neuropsychological assessments were collected for all participants included in the study.

Singaporean Memory Clinic Cohort. The Singaporean cohort included participants from the Memory Clinic at the National University Hospital of Singapore (NUS cohort). The cohort, with an average age of 73.98 ± 8.2 years (51% female), comprised individuals diagnosed with dementia, those with prodromal dementia, and cognitively normal participants. The dataset included 210 cranial CT scans (dementia, n = 104; prodromal dementia, n = 89; cognitively normal, n = 17) and paired MRI scans. Each CT scan had a corresponding paired MRI. CT imaging was performed on a Philips 256 slice scanner (slice collimation: 30×0.625 mm, tube voltage: 120 kVp, tube current modulation: 300 mA reference) at the National University Hospital. Magnetic resonance imaging was performed using a 3 T Magnetom Trio Tim system (Siemens Healthineers AG, Erlangen, Germany) equipped with a 32-channel head coil at the Clinical Imaging Research Center, National University of Singapore. Comprehensive demographic data, medical histories, and clinical evaluations were also collected. Diagnoses of dementia were based on clinical assessments and neuropsychological evaluations.

2.2   Training and Test Data Preparation

The Gothenburg H70 Birth Cohort dataset was used to train the nnUNet and MedNeXT models (Fig. 1). A total of 560 CT scans in axial orientation were used as the primary dataset for training. Segmentation labels for these scans were

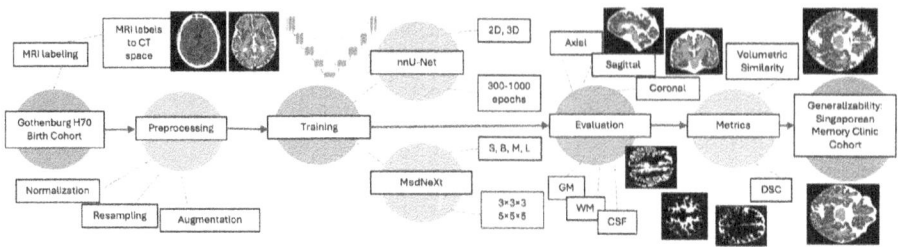

Fig. 1. Diagrammatic representation of the workflow: Preprocessing, training and evaluation of Gothenburg H70 Birth Cohort for nnU-Net (2D, 3D) and MedNeXt (S, B, M, L). Generalizability testing for axial images on the dementia Singaporean Memory Cohort. Gray Matter (GM), White Matter (WM), Cerebrospinal Fluid (CSF) (Color figure online)

derived from MR co-registered segmentation maps and transformed to axial CT space using the statistical parameter mapping software (SPM12). The evaluation involved three distinct test datasets:

- 173 axial CT scans from the Gothenburg H70 Birth cohort were used for model performance comparison between various nnUNet, and MedNeXT models
- 167 sagittal CT scans and 167 coronal CT scans from the Gothenburg H70 Birth Cohort were used for cross-view testing
- 210 axial CT scans from the Singaporean Memory Clinic Cohort were used to evaluate the model's performance in a clinically diverse population

All evaluation test CT datasets had gray matter, white matter, and cerebrospinal fluid labels derived from paired magnetic resonance MRI using the unified segmentation algorithm in SPM12. The paired MRI labels were co-registered with their respective CT spaces in SPM12 and were used as standard criteria for comparison. All input CTs and labels were adapted to the nnU-Net framework with MRI labels merged to a single image with 0 representing background, 1 representing gray matter (GM), 2 representing white matter (WM) and 3 representing cerebrospinal fluid (CSF). The preprocessing included normalization of the intensity values, isotropic resampling to standard voxel spacing for 3D data, and cutting of 3D volumes into individual axial planes for the 2D nnU-Net experiments. Data augmentation techniques, including random rotations, flipping, scaling, and the addition of Gaussian noise, were applied to enhance model robustness during training.

Model Training: NnU-Net. The nnU-Net framework was used to build 2D and 3D convolutional neural network models tailored to segmentation tasks. The 2D nnU-Net was trained on individual axial slices, while the 3D nnU-Net was trained on isotropic 3D volumes. The model-specific configurations were automatically adapted by the nnU-Net preprocessing pipeline, ensuring an optimal architecture and hyperparameters for the characteristics of the dataset. Training experiments were conducted with varying durations to evaluate the impact of extended training. For the 2D and 3D nnU-Net models, axial datasets were trained for both 300 and 1000 epochs. For sagittal and coronal orientations, as well as the dementia cohort, 3D nnU-Net models were uniformly trained for 1000 epochs. All models used the Adam optimizer with an initial learning rate of 0.01, reduced based on the plateauing of the validation performance. Depending on GPU memory constraints, batch sizes were dynamically adjusted between 2 and 16. A composite loss function that combined dice loss and cross-entropy loss was used to optimize segmentation accuracy.

Model Training: MedNeXt. The MedNeXt framework was utilized to build fully ConvNeXt-based 3D segmentation models to address challenges in limited dataset scenarios. MedNeXt integrates a modified 3D-UX-Net architecture, where ConvNeXt blocks replace transformer components, improving gradient flow and maintaining contextual dependencies during upsampling and downsampling. Training experiments were conducted with iterative kernel scaling (UpKern) to address performance saturation in large-kernel networks by initializing them with pretrained smaller kernels. Compound scaling was used to optimize model adaptability by independently tuning the depth, width and size of the network receptive field.

Table 1. MedNeXt training configurations for Kernel=3 and Kernel=5 along with training and validations hours for each model

Model	Kernel size	Task	Parameters	Train(hrs)	Val(hrs)
Small (S)	3×3×3	H70_train	5.6M	48	2
Base (B)	3×3×3	H70_train	10.5M	48	2
Medium (M)	3×3×3	H70_train	17.6M	96	4
Large (L)	3×3×3	H70_train	61.8M	120	6
Large (L)	5×5×5	H70_train	63.0M	288	8

The training configuration included an isotropic 1.0mm spacing and a patch size of 128 × 128 × 128. MedNeXt models were trained with kernels of size 3×3×3 and 5×5×5, for four variants (small, base, medium, large) with parameter counts ranging from 5.6M to 63M. Before training the larger kernel (5×5×5) models, the corresponding 3×3×3 kernel models were pre-trained to initialize weights. We only trained the best performing 3×3×3 kernel for 5×5×5 models due to the lack of computational resources, and all the models were trained for 1000 epochs. Gradient checkpoint was introduced to reduce memory requirements by re-computing activations during backpropagation, enabling efficient training. Training experiments were carried out for all four model sizes with 3×3×3 kernels (small, base, medium and large) and only the large model for

5×5×5 kernels (Table 1). The smaller kernels offered a computationally efficient baseline, while the larger kernels excelled in capturing long-range spatial dependencies crucial for analyzing complex anatomical structures. Inference was performed on 173 datasets for axial, 167 datasets across coronal, and sagittal orientations, with untrained predictions for coronal and sagittal scans providing information on the models' ability to retain long-range dependencies in cases of low spatial resolution. Similarly to nnU-Net (Subsect. 2.2), to optimize the segmentation accuracy, combined dice loss and cross-entry loss were employed.

3   Evaluation

Segmentation performance was assessed using two primary metrics: the Dice Similarity Coefficient (DSC) to measure spatial overlap between predicted segmentation and ground truth, and Pearson's correlation (r) for volumetric similarity to evaluate the alignment of predicted and ground truth volumes. Inference time per sample was also recorded to provide a measure of computational efficiency (Fig. 2).

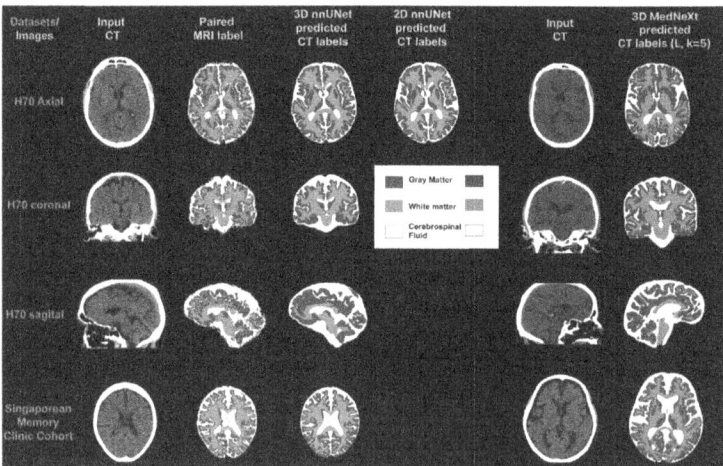

Fig. 2. CT inputs and model outputs: The figure shows input CT images, paired MRI ground truth labels for nnUNet (2D & 3D) and MedNeXt (Large, Kernel = 5) predictions from various test datasets (H70 axial (n = 173), H70 coronal (n = 167), H70 sagittal (n = 167), and Singaporean Memory Clinic Cohort (NUS, n = 210))

The evaluation included several experimental comparisons. *Firstly,* the impact of training duration was assessed by comparing nnUNet trained for 300 and 1000 epochs and for MedNeXt we assessed the duration effects for the small, base, medium and large models using 173 axial test scans. *Secondly,* the generalizability of the 3D nnU-Net and 3D MedNeXt model that performs best (large, k = 5) was tested on sagittal and coronal datasets, quantifying their adaptability to different orientations. Finally, the performance of 3D nnU-Net and 3D MedNext (L, k = 5) trained in 1000 epochs in the dementia cohort was evaluated to determine the robustness and generalizability of the models in a distinct population with potential structural differences in the brain (Fig. 3).

4   Results and Discussion

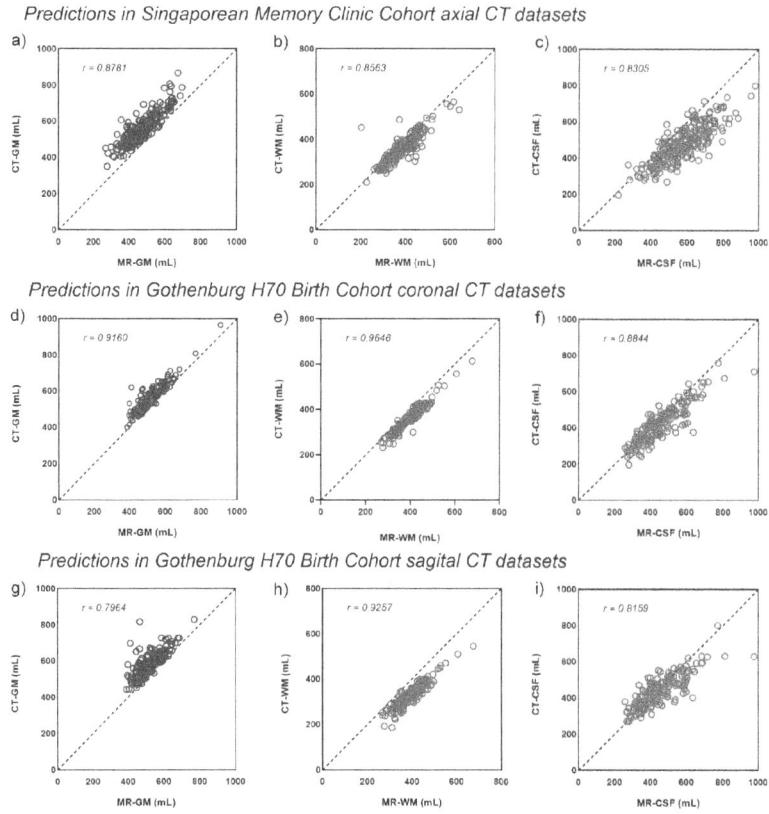

Fig. 3. Correlation plots between 3D nnUNets (trained for 1000 epochs) CT predictions and paired ground truth MRI labels (GM: gray matter, WM: white matter, and CSF: cerebrospinal fluid) in Singaporean Memory Clinic Cohort axial datasets (n = 210; a - c), Gothenburg H70 Birth Cohort coronal (n = 165; d - f), and sagital CT datasets (n = 162; g - i) (Color figure online)

The study demonstrates the potential of MRI-label guided training, using frameworks such as MedNeXt and nnU-Net, for automated brain tissue segmentation in CT imaging. The experiment training was conducted using NVIDIA GPU A30 × 86_64 architecture, 24 GB of VRAM and AMD EPYC 73F3 16-Core

Processor. The training environment included Python 3.11.1, PyTorch version 2.5.1+cu121, nnUNet version 2.5 and MedNext version 1.7.1.

4.1 Inference Time and Training Duration Effects

A total of 717 datasets were used for inference of which 173 were axial CT, 167 sagittal- and coronal CT from the Gothenburg H70 cohort and 210 axial CT from the NUS Dementia cohort. Inference times were consistent across experiments, with an average of 36 s per image for the 2D nnUNet and 47 secs per image for the 3D nnUNet. Although MedNeXt inference times were longer on average, ranging from 50 s per image for small and base models to 3 minutes for the largest model. nnU-Net achieved competitive segmentation accuracy with shorter training and inference times; therefore, it is more feasible for real-time predictions and in settings with limited computational resources.

Table 2. Performance of nnUNet Models Across Cohorts and Configurations

Dataset	Model	Epoch	Dice score coefficient (DSC)			Vol correlation(r)		
			GM	WM	CSF	GM	WM	CSF
H70 A	2D	300	0.79 ± 0.04	0.84 ± 0.03	0.76 ± 0.04	0.941	0.961	0.879
H70 A	2D	1000	0.80 ± 0.04	0.84 ± 0.03	0.77 ± 0.03	0.943	0.965	0.874
H70 A	3D	300	0.80 ± 0.04	0.84 ± 0.04	0.77 ± 0.03	0.892	0.870	0.860
H70 A	3D	1000	$\mathbf{0.81 \pm 0.04}$	$\mathbf{0.86 \pm 0.03}$	$\mathbf{0.78 \pm 0.04}$	0.952	0.970	0.875
H70 C	3D	1000	0.70 ± 0.03	0.76 ± 0.02	0.69 ± 0.04	0.916	0.964	0.884
H70 S	3D	1000	0.74 ± 0.04	0.81 ± 0.03	0.73 ± 0.04	0.796	0.925	0.815
NUS A	3D	1000	0.66 ± 0.04	0.76 ± 0.03	0.68 ± 0.04	0.878	0.856	0.830

The 3D nnU-Net demonstrated a slightly improved performance when trained for 1000 epochs compared to 300 epochs, with the average DSC increasing from 0.80 to 0.81, 0.84 to 0.86, and 0.77 to 0.78, and volumetric correlation improving from 0.89 to 0.95, 0.87 to 0.97, and 0.86 to 0.87, in GM, WM, and CSF respectively. Training duration in 2D nnUNets did not have a significant influence on their performance. In general, 3D nnUNets trained for 1000 epochs performed comparatively better than the other nnUNet models (see Table 2).

The MedNeXt model with the largest configuration (L, k=5) required longer training times due to increase in the number of parameters (see Table 1). The training time scaled significantly with kernel size and network sizes. Small model (3×3×3 kernel) required 48 h for training and the large model (5×5×5 kernel) required 288 h resulting in better performance, with the average DSC of GM, WM and CSF increasing from 0.81, 0.85 and 0.78 to 0.82, 0.87 and 0.79. The performance of the MedNeXt base model was equivalent to the best performing 3D nnUNet model with the DSC of 0.81, 0.86 and 0.78 for GM, WM and CSF respectively (see Table 3) (Figs. 4, 5).

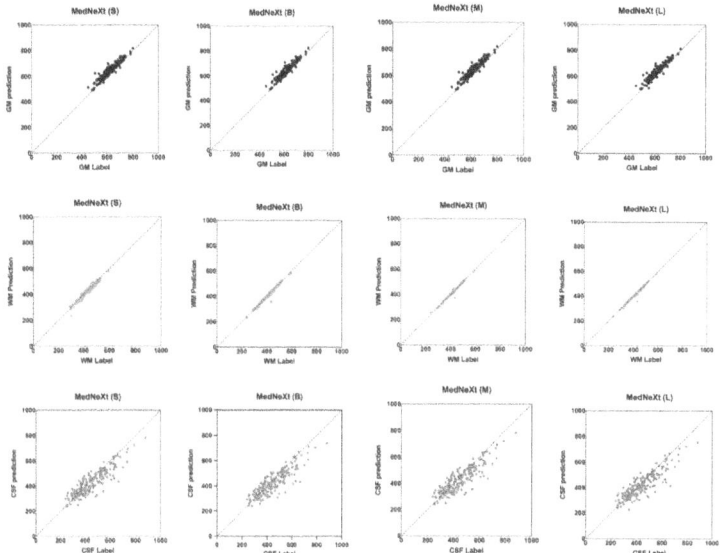

Fig. 4. Correlation plots between 3D MedNeXt (Small, Base, Medium and Large, k=3) axial-CT predictions and paired ground truth MRI labels (GM: gray matter, WM: white matter, and CSF: cerebrospinal fluid) in Gothenburg H70 Birth Cohort (axial, n = 173) (Color figure online)

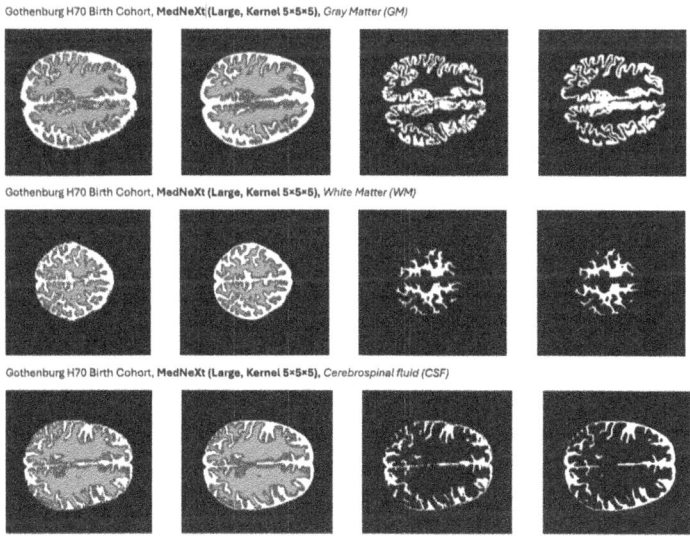

Fig. 5. CT input and model outputs: The figure shows the CT input and MedNeXt predictions from H70 axial dataset (N= 173) for MedNext large (L) model with kernel size of 5×5×5 for GM: gray matter, WM: white matter, and CSF: cerebrospinal fluid (Columns: Input label, Output predictions, Mask and Model (L, k=5))

Table 3. Performance of MedNeXt Models Across Cohorts and Configurations

Dataset	Model	Epoch	Dice score coefficient (DSC)			Vol correlation(r)		
			GM	WM	CSF	GM	WM	CSF
H70 A	S (3)	1000	0.81 ± 0.02	0.85 ± 0.03	0.78 ± 0.02	0.947	0.991	0.874
H70 A	B (3)	1000	0.81 ± 0.04	0.86 ± 0.04	0.78 ± 0.03	0.943	0.991	0.875
H70 A	M (3)	1000	0.82 ± 0.03	0.86 ± 0.02	0.78 ± 0.02	0.947	0.991	0.860
H70 A	L (3)	1000	0.82 ± 0.03	0.86 ± 0.03	0.79 ± 0.02	0.941	0.991	0.870
H70 A	L (5)	1000	**0.82 ± 0.04**	**0.87 ± 0.02**	**0.79 ± 0.04**	0.956	0.992	0.872
H70 C	L (5)	1000	0.71 ± 0.03	0.78 ± 0.02	0.71 ± 0.04	0.928	0.971	0.894
H70 S	L (5)	1000	0.76 ± 0.03	0.83 ± 0.03	0.74 ± 0.02	0.852	0.955	0.842
NUS A	L (5)	1000	0.69 ± 0.04	0.78 ± 0.03	0.71 ± 0.02	0.888	0.991	0.865

4.2   Cross-View Performance

The 3D nnU-Net exhibited good to strong performance across different orientations, achieving average DSCs of 0.82, 0.72, and 0.76 for axial, coronal and sagittal datasets, respectively while the 3d MedNeXt reported better performance with the average DSC of 0.83, 0.73 and 0.78. The volumetric similarity was less across orientations other than the training orientation, with scores ranging from 0.796 (H70 Sagittal- GM) to 0.970 (H70 Axial- WM) for nnUNet and 0.842 (H70 Sagittal- CSF) to 0.992 (H70 Axial WM) for MedNeXt (see Table 2 & Table 3). These evaluations of models across axial, sagittal, and coronal orientations revealed orientation-specific differences in segmentation performance, with axial datasets consistently yielding the highest accuracy. This finding emphasizes the need for orientation-aware considerations when implementing segmentation models in clinical workflows. The results indicate the model's adaptability to different anatomical planes without additional training modifications.

4.3   Generalizability to the Dementia Cohort

The Gothenburg H70 Birth Cohort comprises a homogeneous dataset consisting of acquisitions of CT and MR images from individuals with an average age of 70.44 ± 2.6 years (2.1). However, this homogeneity risks the models inadvertent learning biases due to the lack of diversity within the population sampled. While the training data is gender-balanced (52.6% female), the underrepresentation of certain demographic factors, such as diverse age and ethnicity, could result in misclassification of tissue types for individuals from other demographic groups. To address this limitation, we conducted inference on the NUS Singapore cohort (2.1) to evaluate the model's performance on a dataset representing a different ethnic group, thereby assessing its generalizability. The model's performance on the dementia cohort demonstrated its generalizability, achieving an average DSC and volumetric similarity of 0.70 & 0.854 for nnUNet and 0.73 & 0.912 for MedNeXt (see NUS Axial volumetric correlation, Table 2 and Table 3). MedNeXt consistently outperformed nnU-Net in segmentation accuracy and volumetric similarity across all orientations, illustrating its scalability and robustness. However, improved performance required longer training times and higher computational requirements, which limits its feasibility in resource-constrained environments (Fig. 6).

Singaporean Memory Clinic Cohort, **MedNeXt (Large, Kernel 5×5×5)**, *Gray Matter (GM)*

Singaporean Memory Clinic Cohort, **MedNeXt (Large, Kernel 5×5×5)**, *White Matter (WM)*

Singaporean Memory Clinic Cohort, **MedNeXt (Large, Kernel 5×5×5)**, *Cerebrospinal fluid (CSF)*

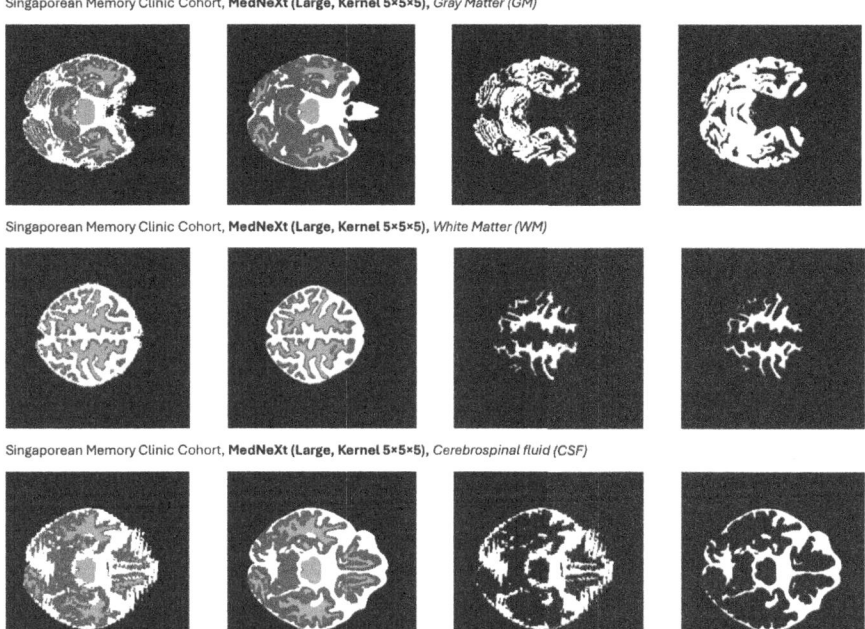

Fig. 6. CT input and model outputs: The figure shows the CT input and MedNeXt predictions from Singaporean Memory Clinic Cohort (N= 210) for MedNext large (L) model with kernel size of 5×5×5 for GM: gray matter, WM: white matter, and CSF: cerebrospinal fluid (Columns: Input label, Output predictions, Mask and Model (L, k=5))

5 Conclusions

The findings reveal that 3D nnU-Net performance improvements observed when increasing epochs from 300 to 1000. However, training duration had minimal impact on 2D nnU-Net models. The MedNeXt model demonstrated superior segmentation accuracy, particularly in cross-orientation generalization and its robustness in the dementia cohort, but with significantly longer training and inference times due to increased model complexity. In cross-view evaluations, MedNeXt consistently outperformed nnU-Net in all anatomical planes, achieving higher DSC and volumetric similarity scores. However, axial datasets consistently produced the highest accuracy in both models, highlighting the need for orientation-sensitive considerations in clinical workflows. When generalizing to the NUS Dementia Cohort, MedNeXt demonstrated better segmentation accuracy and volumetric similarity than nnU-Net across ethnically diverse populations. The computational demands of MedNeXt may limit its feasibility in resource-constrained environments. In terms of efficiency, nnU-Net required significantly shorter training and inference times, which is better for real-time applications .

A   Correlation Plots of nnUNet on H70 Cohort

see Fig. 7 and Fig. 8

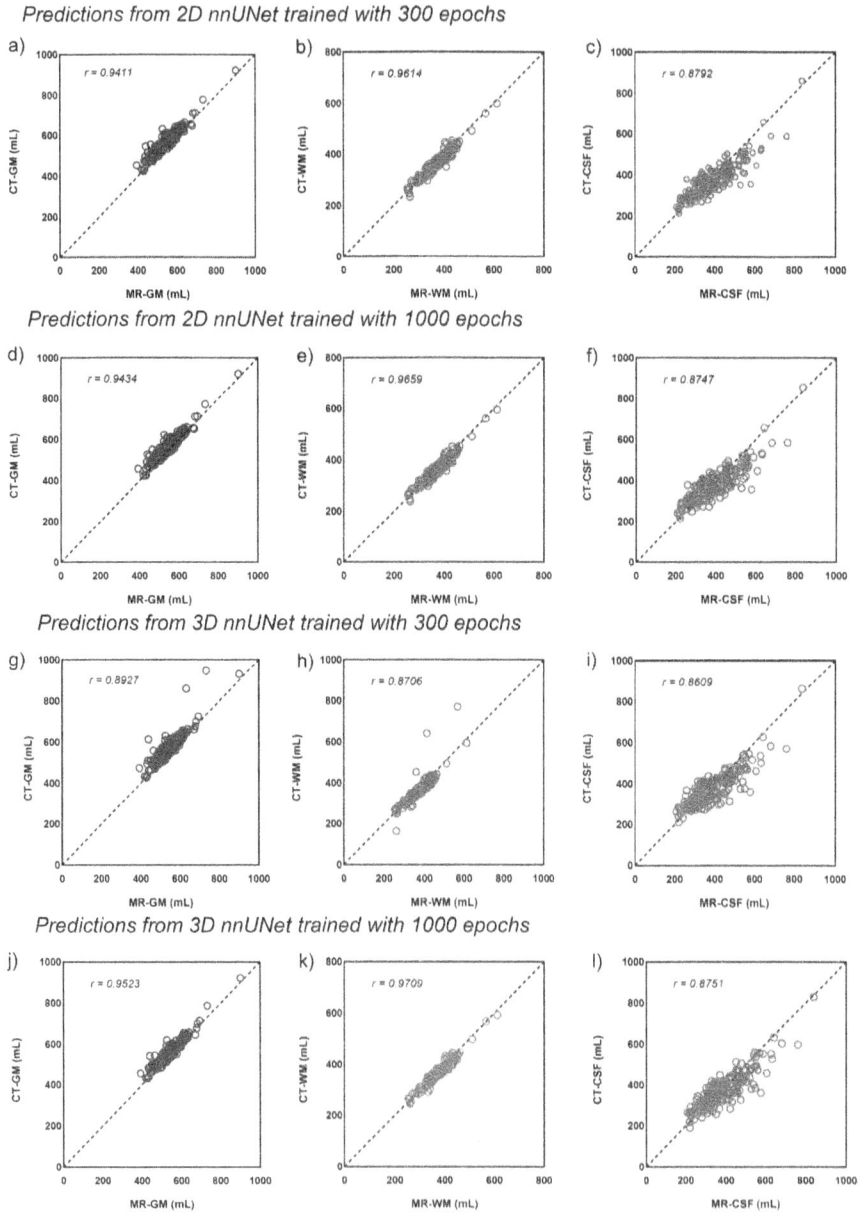

Fig. 7. Correlation plots between 2D (trained for 300 epochs (a - c) and trained for 1000 epochs (d - f) and 3D nnUNets ((trained for 300 epochs (g - i) and trained for 1000 epochs (j - l) CT predictions and paired ground truth MRI labels (GM: gray matter, WM: white matter, and CSF: cerebrospinal fluid) in Gothenburg H70 Birth Cohort axial CT datasets (n = 173) (Color figure online)

B   MedNeXt Prediction for H70 Cohort of GM, WM and CSF for Various Models

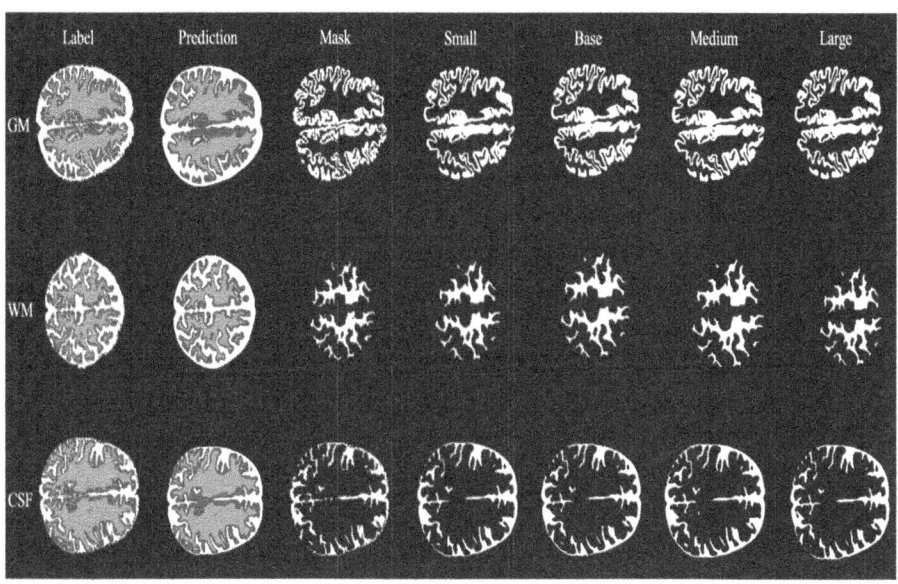

Fig. 8. CT inputs and model outputs: The figure shows paired MRI ground truth labels for MedNeXt predictions (Small, Base, Medium and Large models) from H70 axial dataset of kernel size 3×3×3 for GM: gray matter, WM: white matter, and CSF: cerebrospinal fluid (Color figure online)

References

1. Amarenco, P.: Transient ischemic attack. *New England Journal of Medicine* **382**(20), 1933–1941 (2020)
2. Ashburner, J., et al.: Spm12 manual. Wellcome Trust Centre for Neuroimaging, London, UK, 2464(4) (2014)
3. Despotović, I., Goossens, B., Philips, W.: MRI segmentation of the human brain: challenges, methods, and applications. Comput. Math. Methods Med. **2015**(1), 450341 (2015)
4. Dora, L., Agrawal, S., Panda, R., Abraham, A.: State-of-the-art methods for brain tissue segmentation: a review. IEEE Rev. Biomed. Eng. **10**, 235–249 (2017)
5. González-Villà, S., Oliver, A., Valverde, S., Wang, L., Zwiggelaar, R., Lladó, X.: A review on brain structures segmentation in magnetic resonance imaging. Artif. Intell. Med. **73**, 45–69 (2016)
6. Isensee, F., Jaeger, P.F., Kohl, S.A.A., Petersen, J., Maier-Hein, K.H.: nnU-Net: A self-configuring method for deep learning-based biomedical image segmentation. Nature Methods **18**(2), 203–211 (2021)

7. Islam, M., Vibashan, V. S., Jeya Maria Jose, V., Wijethilake, N., Utkarsh, U., Ren, H.: Brain tumor segmentation and survival prediction using 3D attention U-Net. In: *BrainLes: Glioma, Multiple Sclerosis, Stroke and Traumatic Brain Injuries: 5th International Workshop, BrainLes 2019*, Held in Conjunction with MICCAI 2019, Shenzhen, China, October 17, 2019, Revised Selected Papers, Part I 5, pp. 262–272. Springer (2020)

8. Lee, H.H., Bao, S., Huo, Y., Landman, B.A.: 3D UX-Net: a large kernel volumetric ConvNet modernizing hierarchical transformer for medical image segmentation. arXiv preprint arXiv:2209.15076 (2022)

9. Mavridis, C., et al.: Automatic segmentation in 3D CT images: a comparative study of deep learning architectures for the automatic segmentation of the abdominal aorta. Electronics **13**(24), 4919 (2024)

10. Monteiro, M., et al.: Multiclass semantic segmentation and quantification of traumatic brain injury lesions on head CT using deep learning: an algorithm development and multicenter validation study. The Lancet Digital Health **2**(6), e314–e322 (2020)

11. Rao, B.S., Aparna, M.: A review on Alzheimer's disease through analysis of MRI images using deep learning techniques. IEEE Access (2023)

12. Raza, R., Bajwa, U.I., Mehmood, Y., Anwar, M.W., Jamal, M.H.: DResU-Net: 3D deep residual U-Net based brain tumor segmentation from multimodal MRI. Biomed. Signal Process. Contr. **79**, 103861 (2023)

13. Ronneberger, O., Fischer, P., Brox, T.: U-Net: convolutional networks for biomedical image segmentation. In: Navab, N., Hornegger, J., Wells, W.M., Frangi, A.F. (eds.) MICCAI 2015. LNCS, vol. 9351, pp. 234–241. Springer, Cham (2015). https://doi.org/10.1007/978-3-319-24574-4_28

14. Roy, S., et al.: MedNext: Transformer-driven scaling of ConvNets for medical image segmentation. In: International Conference on Medical Image Computing and Computer-Assisted Intervention, pp. 405–415. Springer (2023)

15. Shamshad, F., et al.: Transformers in medical imaging: a survey. Med. Image Anal. **88**, 102802 (2023)

16. Smith, S.M., et al.: Brain aging comprises many modes of structural and functional change with distinct genetic and biophysical associations. eLife **9**, e52677 (2020)

17. Srikrishna, M., et al.: Deep learning from MRI-derived labels enables automatic brain tissue classification on human brain CT. NeuroImage **244**, 118606 (2021)

18. Yang, Q., et al.: Low-dose CT image denoising using a generative adversarial network with Wasserstein distance and perceptual loss. IEEE Trans. Med. Imag. **37**(6), 1348–1357 (2018)

19. Zhang, Z., Wu, C., Coleman, S., Kerr, D.: Dense-inception U-Net for medical image segmentation. Computer Methods and Programs in Biomedicine **192**, 105395 (2020)

Author Index

© The Editor(s) (if applicable) and The Author(s), under exclusive license
to Springer Nature Singapore Pte Ltd. 2025
S. S. Khan et al. (Eds.): IJCAI 2025, CCIS 2620, p. 147, 2025.
https://doi.org/10.1007/978-981-95-0568-5

The manufacturer's authorised representative in the EU is Springer
Nature Customer Service Centre GmbH, Europaplatz 3, 69115 Heidelberg,
Germany. If you have any concerns regarding our products, please
contact ProductSafety@springernature.com

Printed and bound by CPI Group (UK) Ltd, Croydon, CR0 4YY
06/05/2026
02103601-0006